Data Ethics and Regulatory Concepts in FinTech

Keith M. Walters

Bishop Gate Academic Press

2026

Data Ethics and Regulatory Concepts in FinTech

For permission requests, write to the publisher at:

Bishop Gate Academic Press

Library of Congress Control Number: [Pending]

ISBN 979-8-9951001-3-3 (print)

ISBN 979-8-9951001-2-6 (digital)

First Edition, 2026

Printed in the United States of America

Preface

Financial technology is no longer an emerging sector — it is the sector. The question for anyone working in or entering financial services today is not whether technology will shape their work, but whether they understand the legal, ethical, and governance frameworks that determine what that technology is allowed to do. This book was written to answer that question directly.

Most FinTech education splits across two problems: courses that cover the technology without the regulatory depth, and courses that cover the regulation without grounding it in how the industry actually operates. This text replaces two previously required readings with a single integrated resource built around the conviction that regulation and ethics are not constraints on FinTech — they are the professional foundation on which it stands.

This book was written for working professionals — people who bring real-world experience to their education and who need knowledge they can apply, not theory they have to translate. Whether you are entering the FinTech industry, working within it, or advising firms that operate in it, the demands are the same: understand complex regulatory systems quickly and act on that understanding with confidence. The writing is direct, the examples are current, and the analytical frameworks are practical.

The book moves from the broad regulatory landscape of U.S. FinTech through cryptocurrency, banking, payments, data ethics, privacy law, cybersecurity, risk management, corporate governance, and artificial intelligence. Each chapter builds on what came before. The goal is not to cover every topic in the field — it is to give you the analytical tools to engage with any topic in the field, including ones that do not yet exist.

One final note: this is not a compliance manual. Regulations change, enforcement postures shift, and the technology itself evolves faster than any book can track. What does not change is the need for professionals who understand why the rules exist, not just what they require. That understanding — of the tension between innovation and consumer protection, between data utility and privacy rights, between efficiency and accountability — is what this book is designed to build.

Keith M. Walters

Table of Contents

Chapter 1: The FinTech Landscape — Innovation, Disruption, and Regulation

Introduction

Finance has always been shaped by technology. The introduction of the telegraph allowed banks to transmit wire transfers across the country. The rise of credit cards in the mid-twentieth century changed how consumers spent money. Automated teller machines made cash accessible around the clock. Each of these innovations disrupted existing habits, created new business models, and eventually forced regulators to develop new rules.

What is different today is the pace and breadth of change. The term financial technology — commonly shortened to FinTech — describes an industry that has compressed decades of financial evolution into a few short years. Mobile apps now let consumers open bank accounts, invest in the stock market, send money internationally, and apply for loans, all without setting foot in a branch or speaking to a human being. Blockchain technology enables transactions between strangers on opposite sides of the world without any bank in the middle. Algorithms evaluate creditworthiness in seconds, using data points that traditional lenders never considered.

These changes create enormous opportunity. They also create enormous complexity. Financial services have always been among the most heavily regulated industries in any economy, because the consequences of failure — bank runs, fraud, consumer harm, systemic collapse — are severe. When technology moves faster than regulation, gaps open. And gaps in financial regulation tend to be filled by bad actors, by consumer losses, and eventually by government intervention.

This chapter introduces the FinTech landscape: what it is, what sectors it encompasses, who the players are, and why the tension between innovation and regulation is the defining challenge of the industry. Everything that follows in this textbook — the regulatory frameworks, the ethical dilemmas, the compliance requirements, the enforcement actions — flows from the dynamics introduced here.

Defining FinTech

FinTech is not a single technology or a single industry. It is best understood as the application of technology to improve, automate, or disrupt financial services. That definition is intentionally broad, because FinTech itself is broad. It includes startups trying to replace the traditional bank account and global corporations embedding financial services into platforms people already use. It includes mobile payment apps used by billions and highly specialized software used by a handful of institutional traders.

The word itself — a contraction of financial technology — has been in use since at least the early 1990s, when it referred primarily to the technology systems used by banks and financial institutions to manage their back-office operations. In the modern era, the term has expanded to cover any firm that uses technology to deliver financial services, whether or not that firm is itself a bank. A company does not need a banking license to be a FinTech company. What it needs is a technology-driven approach to solving a financial problem.

This distinction matters for regulation. Traditional banks operate under a comprehensive legal framework built over decades. They hold government-insured deposits, are subject to capital requirements, submit to regular examinations, and operate under the oversight of multiple federal and state agencies. A FinTech company that provides a service similar to banking — accepting funds, making payments, extending credit — may not be subject to the same oversight simply because it does not hold a bank charter. This regulatory asymmetry is one of the central tensions in FinTech policy today.

The Major Sectors of FinTech

FinTech is not monolithic. It is a collection of distinct sectors, each with its own technology stack, business model, customer base, and regulatory environment. Understanding these sectors is essential for understanding the regulatory challenges each one faces.

Payments and Money Transfer

Payments is arguably the oldest and most mature sector of FinTech. Companies like PayPal, founded in 1998, demonstrated that consumers would embrace digital payment alternatives long before smartphones existed. Today, the payments sector includes mobile wallets like Apple Pay and Google Pay, peer-to-peer transfer apps like Venmo and Cash App, business payment platforms like Square and Stripe, and international remittance services like Wise and Remitly.

The common thread is that these companies move money — between individuals, between businesses, or between countries — often faster, cheaper, and more conveniently than traditional banks. Many of them do this by building on top of existing banking infrastructure rather than replacing it. A Venmo transfer, for example, ultimately clears through the Automated Clearing House network, the same system banks have used for decades. Others, particularly in the international remittance space, are developing alternatives that bypass legacy rails entirely.

Payments companies face a web of regulatory requirements: anti-money laundering compliance, consumer protection laws, state-by-state money transmitter licensing, and, increasingly, scrutiny over their handling of disputes and fraud. The Zelle platform — operated by a consortium of major banks — became the subject of Congressional hearings and a Consumer Financial Protection Bureau lawsuit over allegations that it failed to protect consumers from fraud. The case illustrated that even platforms backed by established financial institutions are not immune to regulatory action when consumer harm is at scale.

Lending and Credit

Online lending platforms emerged in the mid-2000s with a straightforward pitch: by connecting borrowers directly with investors and using technology to assess creditworthiness, they could offer lower rates than banks and faster decisions than any human loan officer. LendingClub, founded in 2006, became one of the most prominent examples of this model, known as peer-to-peer lending.

The sector has since evolved well beyond pure peer-to-peer models. Online lenders today include marketplace lenders, balance sheet lenders, buy-now-pay-later providers, and earned wage access platforms. What unites them is the use of technology — often sophisticated algorithms drawing on non-traditional data — to make credit decisions. Some have used social media data, shopping behavior, and even typing patterns to assess borrower risk. These practices raise serious questions about fairness, accuracy, and the potential for algorithmic discrimination.

Lending FinTechs occupy a complicated regulatory space. They are not banks, so they are not subject to bank-level oversight. But they extend credit, so they are subject to federal consumer credit laws and, critically, to state usury laws that cap interest rates. The mismatch between state-by-state rate caps and the desire to operate nationally has driven many online lenders into partnership arrangements with chartered banks — sometimes called rent-a-bank arrangements — that allow them to export more favorable interest rate terms. These arrangements have themselves become a focus of regulatory scrutiny.

Cryptocurrency and Blockchain

When Bitcoin launched in 2009, it was an experiment — a peer-to-peer electronic cash system designed to operate without any central authority. Within a decade, it had spawned an entire ecosystem: thousands of competing cryptocurrencies, blockchain platforms supporting complex financial contracts, and markets valuing the collective asset class at trillions of dollars at its peak.

The crypto sector defies easy categorization. Some cryptocurrencies function as a medium of exchange. Others represent ownership stakes in a platform or project. Still others are tied to the value of a traditional asset, like the U.S. dollar, and are used primarily as a stable store of value within crypto markets. The technology underlying these assets — distributed ledger technology, or blockchain — has applications well beyond digital currency, from supply chain tracking to smart contracts that execute automatically when conditions are met.

Regulators have struggled to classify and oversee this sector. Is a cryptocurrency a currency, a commodity, or a security? The answer determines which agency has jurisdiction and what rules apply — and different U.S. agencies have offered different answers. The Securities and Exchange Commission has pursued enforcement actions against crypto companies it views as having sold unregistered securities. The Commodity Futures Trading Commission has asserted jurisdiction over others. The collapse of the FTX exchange in late 2022, which wiped out billions in customer funds and resulted in criminal convictions for its founder, became the defining example of what happens when a sector grows faster than its regulatory framework.

WealthTech and Investment Platforms

Technology has also transformed how people invest and manage money. Robo-advisors — platforms that use algorithms to build and manage investment portfolios — emerged in the years following the 2008 financial crisis as a low-cost alternative to human financial advisers. Betterment and Wealthfront were among the first, offering automated portfolio management for fees far below what a traditional adviser would charge. Today, robo-advisory features are offered by traditional brokerages as well, blurring the line between old and new.

At the retail end of the spectrum, platforms like Robinhood democratized stock trading by eliminating commissions and offering a sleek mobile interface. The consequences of that democratization became apparent during the GameStop trading frenzy of early 2021, when Robinhood's decision to restrict purchases of certain stocks — citing clearing house deposit requirements — triggered a firestorm of criticism from retail investors and drew scrutiny from Congress and regulators. The episode raised fundamental questions about the obligations of trading platforms, the risks of gamification in investing, and the adequacy of existing investor protection frameworks.

InsurTech and RegTech

Two additional sectors are worth naming here, though they receive less public attention than payments or crypto. InsurTech applies technology to the insurance industry — using data analytics, telematics, and artificial intelligence to price policies more precisely, streamline claims processing, and offer coverage products that traditional insurers have been slow to develop. RegTech — regulatory technology — is the application of technology to compliance itself. RegTech firms help financial institutions manage their regulatory obligations, automate reporting, monitor transactions for suspicious activity, and maintain the documentation regulators require. In a regulatory environment as complex as the one facing FinTech, the demand for RegTech solutions has grown substantially.

The Key Players

Understanding the FinTech landscape requires understanding who the players are and how they relate to each other. The ecosystem is more complex than a simple story of startups disrupting banks.

FinTech Startups

The popular image of FinTech is a startup: a small team with a disruptive idea, venture capital funding, and no legacy technology or regulatory obligations holding it back. This image captures something real. Startups have driven genuine innovation in financial services, forcing incumbents to adapt and expanding access to financial products for underserved consumers. But startups also face significant disadvantages. They lack the customer base, brand recognition, regulatory relationships, and capital reserves of established institutions. They often discover, sometimes painfully, that scaling a financial service means acquiring exactly the compliance infrastructure and regulatory obligations they hoped to avoid.

Traditional Financial Institutions

Banks, credit card companies, brokerages, and insurers are not passive bystanders in the FinTech revolution. Many have invested heavily in technology modernization, launched their own digital products, and acquired or partnered with FinTech startups. JPMorgan Chase, Bank of America, and other major banks have spent billions on technology in recent years. The line between a bank and a FinTech company is increasingly blurry. What distinguishes traditional institutions is not their lack of technology but their regulatory status — they hold charters, carry deposit insurance, and operate under comprehensive oversight frameworks built over decades.

The relationship between banks and FinTechs is not simply competitive. It is often collaborative. Many FinTech companies rely on bank partnerships to access the payment systems, deposit infrastructure, and regulatory permissions they need to operate. Banks, in turn, benefit from the technology capabilities and customer acquisition prowess of their FinTech partners. This interdependence shapes the regulatory landscape in important ways.

Regulators

The United States does not have a single financial regulator. It has many, operating at both the federal and state levels, each with jurisdiction over specific types of institutions or activities. The Office of the Comptroller of the Currency charters and supervises national banks. The Federal Reserve oversees bank holding companies and plays a central role in payment system oversight. The Federal Deposit Insurance Corporation insures deposits and regulates state-chartered banks that are not members of the Federal Reserve System. The Securities and Exchange Commission regulates securities markets and investment products. The Commodity Futures Trading Commission oversees derivatives markets. The Consumer Financial Protection Bureau enforces federal consumer financial laws. And each of the fifty states maintains its own regulatory framework for banking, lending, money transmission, and insurance.

For a FinTech company trying to operate nationally, this fragmented landscape presents a significant challenge. A payments company may need money transmitter licenses in every state where it operates. A lending platform must navigate a patchwork of state interest rate caps. A crypto exchange faces overlapping and sometimes contradictory claims of jurisdiction from multiple federal agencies. The cost and complexity of compliance is itself a competitive factor in the industry.

The Core Tension: Innovation and Regulation

Every chapter in this textbook returns, in some form, to the same fundamental tension: the drive to innovate and the need to regulate. It is worth understanding this tension clearly from the outset, because it is easy to caricature it as a simple conflict between progress and bureaucracy. The reality is more nuanced.

Financial regulation exists for good reasons. The history of finance is a history of crises, frauds, and failures that have caused real harm to real people. The Great Depression demonstrated what happens when banks fail without deposit insurance. The savings and loan crisis of the 1980s demonstrated what happens when deregulation outpaces prudential oversight. The 2008 financial crisis demonstrated what happens when complex financial instruments proliferate faster than the understanding of their risks. Regulation is the accumulated response to these lessons. It is not perfect, and it is not always timely, but it serves purposes that matter: protecting consumers, maintaining confidence in financial systems, preventing fraud, and reducing the risk that a single failure cascades into systemic collapse.

At the same time, regulation can stifle innovation that would genuinely benefit consumers and the broader economy. Overly prescriptive rules written for one era of technology may make no sense in the next. Licensing requirements that were designed to screen out bad actors can also screen out legitimate innovators who lack the resources to navigate a complex compliance process. Regulatory uncertainty — not knowing whether a new product will be treated as a security, a money transmitter service, or something else entirely — can prevent beneficial innovations from reaching the market at all.

The challenge for policymakers is to design regulatory frameworks that are robust enough to prevent harm and flexible enough to accommodate innovation. The challenge for FinTech companies is to build businesses that can survive in that environment — which means understanding regulation not as an obstacle to be avoided but as a constraint to be navigated thoughtfully. The challenge for professionals working in this industry, and for students preparing to enter it, is to understand both dimensions: what the regulations require and why they exist.

Why FinTech Complicates Existing Frameworks

Traditional financial regulation was built around institutions: banks, broker-dealers, insurance companies. The regulatory framework attached to the institution. A company that held a bank charter was subject to bank regulation; a company that did not was not. FinTech disrupts this logic by enabling firms to deliver services that look like banking, investing, or insurance without holding the charters that traditionally triggered regulatory oversight.

Consider a mobile app that allows users to deposit funds, earn interest, and make payments. Is that a bank? If it holds deposits and the funds are insured, it may be. If it partners with a bank that holds the deposits while the app handles the customer interface, it may not be — even though the consumer experience is essentially the same. Whether that distinction should matter, and how regulators should treat it, is a live policy debate with significant implications for consumer protection, competitive fairness, and the future structure of financial services.

The pace of technological change compounds the problem. Regulators typically move slowly and deliberately — drafting rules, soliciting public comment, finalizing guidance, and enforcing over time. Technology moves in the opposite direction. By the time a regulatory framework for one innovation is in place, several new ones may have emerged. This lag is not unique to financial regulation, but its consequences in finance are particularly serious, because the assets at stake — people's savings, retirement funds, and financial security — are not easily recovered once lost.

The Global Dimension

FinTech is a global industry operating in a world of national regulatory systems. A crypto exchange can be headquartered in one country, incorporated in another, and serve customers in dozens more. This creates both opportunity and risk. Firms can seek out jurisdictions with lighter regulation — a practice sometimes called regulatory arbitrage — and use that advantage to serve customers in markets with stricter rules. Regulators in one country have limited ability to police firms based elsewhere.

The United States is not the only major player in FinTech regulation, and comparisons to other regulatory approaches are instructive. The United Kingdom's Financial Conduct Authority developed one of the first regulatory sandboxes — controlled environments where firms can test new products under relaxed rules before full regulatory compliance is required. This approach has been influential globally and has been replicated in various forms in Singapore, Australia, and several U.S. states. The European Union's General Data Protection Regulation has set a global benchmark for data privacy that affects any firm serving European customers, including U.S.-based FinTechs. These international developments matter for U.S. professionals, not because they govern domestic operations, but because they shape the competitive and regulatory context in which global FinTech operates.

Key Drivers of FinTech Growth

FinTech's rapid growth is not accidental. Several converging forces have created the conditions for the industry to flourish.

Smartphones put powerful computing devices in the hands of billions of people, creating a distribution channel for financial services that required no physical infrastructure. The 2008 financial crisis damaged public trust in traditional banks and created an opening for alternatives. Low interest rates in the years following the crisis pushed investors toward higher-yielding assets and made venture capital funding available at scale. A generation of consumers who grew up with digital services expected their financial products to work the same way — immediate, mobile, and transparent. Open banking initiatives, which allow third-party developers to access bank data with customer permission, created technical pathways for new products to be built on top of existing financial infrastructure.

Advances in artificial intelligence and machine learning have accelerated the pace further. Tasks that once required human judgment — evaluating a loan application, detecting fraud, providing investment advice — can now be automated at scale. The volume of data available to train these systems has grown enormously, as financial transactions, social behavior, and consumer activity are recorded and stored in ever-increasing detail. The result is a competitive environment where the most sophisticated data capabilities translate directly into product and pricing advantages.

These drivers also create new vulnerabilities. The same digital channels that make financial services more accessible also make them more susceptible to cybersecurity threats. The same data that enables personalized services also creates privacy risks. The same algorithms that improve efficiency can embed and amplify bias. Understanding these trade-offs is not optional for professionals in this field — it is central to doing the job well.

Looking Ahead: What This Textbook Covers

This chapter has introduced the FinTech landscape in broad strokes. The chapters that follow develop each dimension in depth.

Chapters 2 through 6 focus on the regulatory environment: the agencies, frameworks, and specific rules that govern FinTech operations in the United States. Chapter 2 maps the U.S. regulatory landscape in detail, covering the major agencies, the challenge of regulatory fragmentation, and the tools regulators have developed — including sandboxes and innovation offices — to engage with FinTech firms. Chapters 3 and 4 turn to cryptocurrency, covering how it works, how it is classified under existing law, and how regulators have responded to the sector's growth and its failures. Chapters 5 and 6 address payments, banking, and wealth management, tracing how FinTech has challenged legacy systems and what the regulatory implications of that challenge have been.

Chapters 7 and 8 shift to data ethics and privacy law. They examine the ethical questions raised by the collection, use, and monetization of consumer data, and the legal frameworks — including U.S. privacy law and international benchmarks — that govern those activities.

Chapters 9 and 10 address cybersecurity: the threats facing FinTech firms, the defenses available, and the legal and policy landscape governing cybersecurity obligations, with particular attention to the Cybersecurity Information Sharing Act of 2015.

Chapters 11 and 12 cover risk management, internal controls, corporate governance, and the anti-money laundering and know-your-customer frameworks that are foundational to compliance in financial services.

Chapter 13 closes the book with an examination of artificial intelligence in FinTech — its applications, its risks, the problem of algorithmic bias, and the emerging regulatory response.

Throughout, the goal is the same: to give you the knowledge and the analytical tools to understand what is happening in this industry, why it matters, and how to navigate it as a professional.

Key Takeaways

FinTech refers to the application of technology to improve, automate, or disrupt financial services. It encompasses a wide range of sectors, including payments, lending, cryptocurrency, investment platforms, insurance, and regulatory technology.

The FinTech ecosystem involves startups, traditional financial institutions, and regulators, whose relationships are often as collaborative as they are competitive. No single U.S. agency oversees FinTech; instead, multiple federal and state regulators share jurisdiction based on the type of activity involved.

The defining tension of the FinTech industry is the conflict between the pace of technological innovation and the slower, more deliberate pace of regulatory development. This tension is not simply about progress versus bureaucracy — it reflects genuine trade-offs between enabling beneficial innovation and protecting consumers and financial stability.

FinTech complicates traditional regulatory frameworks because regulation was designed around institutions, while FinTech often enables firms to deliver financial services without holding the charters that traditionally triggered regulatory oversight. Managing this challenge is the central task of FinTech regulation, and understanding it is the central purpose of this textbook.

Chapter 2: The U.S. Regulatory Framework for FinTech

Introduction

The United States has one of the most complex financial regulatory systems in the world. That complexity is not the product of poor planning — it reflects the country's history, its federal structure, and deliberate policy choices made over more than a century of financial crises, legislative responses, and institutional evolution. For a traditional bank, this complexity is manageable, if burdensome. For a FinTech company trying to scale a new product nationally, it can be paralyzing.

Understanding who regulates what — and why — is not a bureaucratic exercise. It is a practical necessity for anyone working in or around the FinTech industry. The wrong assumption about which agency has jurisdiction over a particular product can expose a company to enforcement action, fines, or forced shutdown. The right understanding of the regulatory landscape can help a company find pathways to operate legally, engage proactively with regulators, and build the compliance infrastructure that investors, partners, and customers increasingly demand.

This chapter maps the U.S. regulatory framework as it applies to FinTech. It covers the major federal and state agencies, the challenge of regulatory fragmentation, the innovation tools regulators have developed to engage with FinTech firms, and the OCC's special purpose charter pathway for companies seeking bank status. It also situates the U.S. approach in a broader international context, because the choices other countries have made offer useful comparisons — and sometimes competitive pressure — for U.S. policy.

One principle runs through everything in this chapter: there is no one-size-fits-all approach to FinTech regulation. The regulatory treatment of any given company depends on what it does, how it does it, who its customers are, and where it operates. That variability is a feature of the system — it allows rules tailored to specific risks — but it is also a source of enormous complexity for firms trying to navigate it.

The Landscape of U.S. Financial Regulation

Before examining how specific agencies engage with FinTech, it helps to understand the broader architecture of U.S. financial regulation. That architecture has three defining characteristics: it is fragmented across multiple federal agencies; it is divided between federal and state authority; and it is organized primarily around the type of institution rather than the type of activity.

Fragmentation at the federal level means that no single agency has comprehensive oversight of the financial system. Different agencies have jurisdiction based on whether a firm is a bank, a broker-dealer, a futures commission merchant, a money transmitter, or a consumer lender — and those categories do not always map cleanly onto what FinTech firms actually do. A company that accepts customer funds, invests them, and allows customers to make payments might simultaneously implicate the jurisdiction of the OCC, the SEC, the CFPB, and FinCEN, depending on the specific structure of its product.

The federal-state divide adds another layer. State governments have traditionally regulated banking, insurance, and securities within their borders, and they retain significant authority over activities like money transmission and consumer lending. This means a FinTech company that wants to operate in all fifty states may need to obtain licenses in each of them — a process that is time-consuming, expensive, and inconsistent from state to state.

The institution-based structure of regulation creates the most fundamental challenge for FinTech. Rules written for banks apply to banks. Rules written for broker-dealers apply to broker-dealers. A

FinTech company that does something that looks like banking, or investing, or lending, but does not fit into any established institutional category, may find itself in a regulatory gray zone — neither clearly subject to existing rules nor clearly exempt from them. That uncertainty creates risk for the firm and, ultimately, for the consumers it serves.

The Major Federal Regulatory Agencies

The following agencies are the primary federal regulators whose jurisdiction touches FinTech. Each has a distinct mandate, a distinct set of tools, and a distinct relationship with the firms it oversees. The table below provides a summary reference, followed by a more detailed treatment of each agency's role.

Agency	Full Name	Primary FinTech Jurisdiction
OCC	Office of the Comptroller of the Currency	National banks; FinTech charter applicants
Federal Reserve	Board of Governors of the Federal Reserve System	Bank holding companies; payment systems oversight
FDIC	Federal Deposit Insurance Corporation	State-chartered non-member banks; deposit insurance
SEC	Securities and Exchange Commission	Securities offerings; investment platforms; crypto assets deemed securities
CFTC	Commodity Futures Trading Commission	Derivatives; crypto assets deemed commodities
CFPB	Consumer Financial Protection Bureau	Federal consumer financial protection laws; non-bank FinTech lenders and servicers
FinCEN	Financial Crimes Enforcement Network	Anti-money laundering; Bank Secrecy Act compliance; money transmitters
FTC	Federal Trade Commission	Unfair or deceptive practices; data privacy enforcement
State Regulators	State banking, securities, and insurance departments	Money transmitter licensing; state consumer lending laws; state securities laws

The Office of the Comptroller of the Currency

The Office of the Comptroller of the Currency, or OCC, is an independent bureau within the U.S. Department of the Treasury. It charters, regulates, and supervises national banks — banks that operate under a federal charter rather than a state charter. In practical terms, this means the OCC sets the rules for the largest and most prominent banking institutions in the country, including JPMorgan Chase, Bank of America, and Wells Fargo.

For FinTech companies, the OCC is significant for two reasons. First, many FinTechs partner with OCC-regulated banks to access banking infrastructure, deposit accounts, and payment systems. The OCC's expectations for how banks manage those partnerships — including due diligence requirements — shape the terms and timelines of bank-FinTech relationships. Second, the OCC has proposed a pathway for FinTech firms to obtain a special purpose national bank charter, which would allow them to operate as banks without accepting deposits. That proposal, and its implications, is discussed in detail later in this chapter.

In 2016, the OCC established the Office of Innovation, which serves as a central point of contact for FinTech companies seeking guidance on regulatory matters. The Office of Innovation offers responsible innovation framework guidance, conducts outreach to the FinTech community, and facilitates conversations between emerging firms and OCC staff before formal licensing or application processes begin. This was a significant signal that the OCC recognized FinTech as a distinct regulatory challenge requiring dedicated attention rather than treatment as a subset of ordinary bank supervision.

The Federal Reserve

The Federal Reserve System is the central bank of the United States. Its regulatory responsibilities in the FinTech context fall into two main areas: oversight of bank holding companies and their subsidiaries, and oversight of the payment systems that underlie most financial transactions in the country.

A FinTech company that partners with or acquires a bank may find itself subject to Federal Reserve oversight if the transaction creates a bank holding company structure. The Federal Reserve also has authority over state-chartered banks that are members of the Federal Reserve System — a category that includes many of the country's mid-sized and regional banks that serve as partners for FinTech firms.

On the payments side, the Federal Reserve operates the Fedwire Funds Service, one of the primary large-value interbank payment systems in the country, and the FedACH system, which processes the majority of electronic retail payments. In 2023, the Federal Reserve launched FedNow, a real-time payment system designed to enable instant transfers between bank accounts. FedNow represents the Federal Reserve's direct response to the competitive pressure that FinTech payment platforms have placed on the traditional banking system — and its entry into real-time payments has significant implications for the competitive dynamics of the payments sector.

The Federal Deposit Insurance Corporation

The Federal Deposit Insurance Corporation, or FDIC, was created in 1933 in response to the wave of bank failures during the Great Depression. Its primary mission is to maintain stability and public confidence in the banking system by insuring deposits at member institutions. Deposit insurance — currently covering up to $250,000 per depositor per institution — is the reason that most Americans do not worry about losing their savings when a bank fails.

The FDIC also serves as the primary federal regulator for state-chartered banks that are not members of the Federal Reserve System. This gives it supervisory authority over a significant portion of the banking sector, including many of the community banks that FinTech companies have partnered with to offer banking-adjacent services.

For FinTech firms, the FDIC's most direct relevance is through the industrial loan company, or ILC, charter. ILCs are state-chartered financial institutions that can accept federally insured deposits but are not subject to Federal Reserve oversight as bank holding companies. Several major corporations — including Target and Walmart — have applied for ILC charters to gain banking capabilities without submitting their entire corporate structure to bank holding company regulation. Walmart's 2005 ILC application was ultimately denied following significant opposition from the banking industry and consumer advocates. The ILC pathway remains a live option for some FinTech firms, though it is narrow, politically contentious, and subject to FDIC approval.

The Securities and Exchange Commission

The Securities and Exchange Commission, or SEC, was created by the Securities Exchange Act of 1934, in the aftermath of the stock market crash of 1929 and the frauds that contributed to it. The SEC's mandate is to protect investors, maintain fair and orderly markets, and facilitate capital formation. It does this by requiring disclosure from companies that sell securities to the public, regulating the markets and intermediaries through which securities are traded, and enforcing the securities laws.

In the FinTech context, the SEC's most prominent role has been in the regulation of cryptocurrency. The central question is whether a given crypto asset constitutes a security under federal law. If it does, it must be registered with the SEC or qualify for an exemption, and the platforms that trade it must register as exchanges or broker-dealers. The SEC has taken the position that many crypto assets — particularly those sold to investors with the expectation of profit from the efforts of others — are securities, and it has pursued enforcement actions against numerous crypto companies on that basis.

The SEC also regulates investment platforms and robo-advisors. A company that provides investment advice for compensation is generally required to register as an investment adviser under the Investment Advisers Act of 1940. Robo-advisors have generally complied with this requirement, though the application of fiduciary duty obligations to algorithmic advice has raised interpretive questions that the SEC has addressed through guidance and rulemaking.

The SEC has established its own Office of Innovation and Financial Technology, which provides guidance to FinTech firms on registration requirements, no-action requests, and other regulatory questions. Like the OCC's Office of Innovation, it represents an acknowledgment that FinTech raises novel regulatory questions that benefit from dedicated engagement.

The Commodity Futures Trading Commission

The Commodity Futures Trading Commission, or CFTC, regulates the U.S. derivatives markets — futures, options, and swaps. Its authority is defined primarily by the Commodity Exchange Act, which gives it jurisdiction over trading in commodity interests.

The CFTC has asserted that several major cryptocurrencies, including Bitcoin and Ethereum, are commodities rather than securities, placing them within its jurisdiction. This has created a jurisdictional tension with the SEC, which has taken a different view of certain crypto assets. The overlap between CFTC and SEC authority over crypto has been one of the most contested regulatory questions in the sector and has contributed to the uncertainty that FinTech companies operating in the crypto space face.

A FinTech company involved in cryptocurrency derivatives — futures contracts or options on crypto assets — would require registration with the CFTC as a designated contract market, swap execution facility, or futures commission merchant, depending on the specific nature of its activities.

The Consumer Financial Protection Bureau

The Consumer Financial Protection Bureau, or CFPB, was created by the Dodd-Frank Wall Street Reform and Consumer Protection Act of 2010, in direct response to the consumer protection failures that contributed to the 2008 financial crisis. Its mandate is straightforward: protect consumers in the market for consumer financial products and services. It does this through rulemaking, supervision, and enforcement of federal consumer financial laws.

The CFPB's significance for FinTech is hard to overstate. Unlike most federal financial regulators, the CFPB has explicit authority over non-bank financial service providers — which means it can supervise and take enforcement action against FinTech companies that offer consumer financial products, even if those companies are not banks. This authority extends to mortgage lenders, payday lenders, student loan servicers, money transmitters, and others that operate outside the traditional banking charter system.

The CFPB has used this authority actively. It has taken enforcement actions against online lenders for unfair and deceptive practices, sued operators of payment platforms for failure to protect consumers from fraud, and issued rules affecting the earned wage access and buy-now-pay-later sectors. Its rulemaking under Section 1033 of Dodd-Frank — which requires financial institutions to share consumer financial data with authorized third parties — has significant implications for the open banking ecosystem on which many FinTech products depend.

The CFPB has also developed tools specifically designed to engage with FinTech innovation. Project Catalyst is the bureau's initiative for encouraging consumer-friendly innovation in financial services. Through Project Catalyst, the CFPB conducts research on emerging FinTech products, engages with innovators to understand new business models, and issues guidance to help companies understand how consumer protection requirements apply to novel products.

Among the most practically significant tools the CFPB offers is the No-Action Letter. A No-Action Letter is a formal statement from the CFPB indicating that it does not intend to recommend supervisory or enforcement action against a company for a particular product or service, provided the company offers it in the way described in its request. No-Action Letters do not grant legal immunity, but they offer a meaningful degree of regulatory certainty to companies operating in gray areas — and that certainty has real value for FinTech firms trying to determine whether a novel product will expose them to regulatory risk.

The CFPB has faced significant political and legal controversy in recent years. Legal challenges to its funding structure reached the Supreme Court, and shifting administrations have produced substantial changes in its enforcement priorities and staffing. As of early 2026, the bureau's role and scope remain subjects of active political debate. For FinTech professionals, this uncertainty about the CFPB's future direction is itself a regulatory risk to be managed.

The Financial Crimes Enforcement Network

The Financial Crimes Enforcement Network, or FinCEN, is a bureau of the U.S. Department of the Treasury. Its mission is to safeguard the financial system from illicit use, combat money laundering, and promote national security through the collection, analysis, and dissemination of financial intelligence. FinCEN administers the Bank Secrecy Act, which requires financial institutions to assist government agencies in detecting and preventing money laundering.

For FinTech companies, FinCEN's most direct impact is through the money transmitter framework. Under the Bank Secrecy Act, any business that transmits money — moving funds on behalf of customers — is classified as a money services business and must register with FinCEN, implement anti-money laundering programs, file suspicious activity reports, and maintain records of transactions above certain thresholds.

FinCEN has clarified that this framework applies to virtual currency businesses. Companies that administer or exchange virtual currencies are classified as money services businesses and must comply with Bank Secrecy Act requirements. This applies whether the company holds customer funds in a centralized account or facilitates the exchange of cryptocurrency between parties. The know-your-customer requirements that flow from Bank Secrecy Act compliance — verifying the identity of customers, understanding the nature of their transactions, and monitoring for suspicious activity — are among the most operationally demanding compliance obligations that FinTech companies face.

The Federal Trade Commission

The Federal Trade Commission, or FTC, enforces federal consumer protection laws that prohibit unfair or deceptive acts or practices in commerce. It is not a financial sector regulator per se, but its authority extends to FinTech companies that engage in deceptive marketing, fail to honor privacy commitments to consumers, or engage in unfair data practices.

The FTC's authority over data privacy is particularly relevant for FinTech, given that data collection and use is central to most FinTech business models. The FTC has brought enforcement actions against companies for misrepresenting how they collect and use consumer data, failing to implement reasonable data security practices, and sharing consumer data in ways inconsistent with their stated privacy policies. These enforcement actions — backed by the FTC's ability to seek civil penalties and require compliance programs — give the agency meaningful leverage over the data practices of non-bank FinTech firms.

State Regulation and the Challenge of Fragmentation

Federal regulation, substantial as it is, represents only part of the regulatory landscape facing FinTech companies. State governments retain significant authority over financial activities within their borders, and for FinTech companies seeking to operate nationally, navigating fifty separate state regulatory regimes is one of the most significant operational challenges they face.

Money Transmitter Licensing

The most immediate manifestation of state regulatory burden for payments FinTechs is the money transmitter licensing requirement. Most states require companies that transmit money on behalf of customers to obtain a state money transmitter license before operating. The requirements for obtaining these licenses — minimum net worth, surety bonds, background checks, application fees, and ongoing reporting obligations — vary significantly from state to state.

For a FinTech company that wants to offer payment services nationally, this means obtaining and maintaining licenses in up to fifty jurisdictions, each with its own application process, renewal schedule, and compliance requirements. The cost and administrative burden of this process is substantial. It has driven some companies to limit their initial geographic footprint, others to seek partnerships with already-licensed entities, and others to advocate for regulatory reform.

Some states have taken steps to streamline the process. The Money Transmission Modernization Act, which has been adopted in a number of states, is an effort to harmonize money transmitter requirements across state lines and reduce the compliance burden for companies operating in multiple jurisdictions. Progress has been uneven, and the patchwork remains a defining feature of the regulatory landscape for payments FinTechs.

State Consumer Lending Laws

State usury laws — laws that cap the interest rates that lenders can charge — create a parallel challenge for lending FinTechs. Interest rate caps vary significantly across states. A rate that is lawful in one state may be illegal in another. For an online lender that wants to offer a uniform product to customers nationwide, this creates a fundamental problem: it cannot comply with all fifty state rate caps simultaneously while maintaining a single product.

This tension has driven many online lenders into partnerships with chartered banks. Under the so-called valid-when-made doctrine, a loan that is lawful when originated by a bank retains its lawful status even when sold to a non-bank investor. This allows a FinTech company to originate loans through a bank partner at rates that might exceed state caps, then purchase or service those loans as a non-bank entity. These arrangements — sometimes called rent-a-bank structures — have generated regulatory and legal controversy, with some state regulators arguing that the structures are designed to evade consumer protection laws.

State Securities Regulation

State securities regulators — operating under the authority of state blue sky laws — have jurisdiction over the offer and sale of securities within their borders. For FinTech companies that offer investment products, initial coin offerings, or other instruments that may qualify as securities, this adds another layer of compliance complexity on top of SEC registration requirements. State securities regulators have been active in bringing enforcement actions against crypto companies, particularly those that conducted token sales without registering under state law.

Why Fragmentation Is a Defining Challenge

The cumulative effect of federal agency fragmentation and state-level regulatory variation is a compliance environment that is uniquely demanding in the United States compared to most other developed economies. In the United Kingdom, a single regulator — the Financial Conduct Authority — has comprehensive authority over most financial services activities. In Singapore, the Monetary Authority of Singapore serves a similar unified function. In the United States, a FinTech company may simultaneously be subject to OCC guidance on its bank partnership, CFPB supervision of its consumer lending practices, FinCEN registration for its payment activities, SEC scrutiny of any investment products it offers, FTC enforcement of its data practices, and money transmitter licensing in each of the states where it operates.

This is not an argument that the U.S. system is wrong. The fragmented structure reflects deliberate choices about federalism, checks and balances, and the risks of concentrating too much authority in a single regulatory body. But it is essential for FinTech professionals to understand that fragmentation is not simply a technical inconvenience — it is a structural feature of the U.S. system that shapes business strategy, compliance costs, and competitive dynamics in the industry.

Regulatory Innovation Tools

Regulators have not been passive in the face of the FinTech challenge. Over the past decade, several agencies have developed tools specifically designed to engage with innovative firms, provide regulatory clarity, and facilitate responsible experimentation. These tools vary in scope and legal effect, but they share a common purpose: creating pathways for new entrants to navigate the regulatory system without simply leaving them to figure it out on their own.

Regulatory Sandboxes

A regulatory sandbox is a controlled environment in which companies can test new financial products or services under relaxed regulatory requirements, typically for a defined period and subject to conditions designed to protect consumers. The sandbox concept allows regulators and innovators to learn from each other — regulators gain insight into how new products work and what risks they create, while companies gain the ability to test whether their product works in practice before investing fully in regulatory compliance.

The idea originated in the United Kingdom, where the Financial Conduct Authority launched one of the world's first regulatory sandboxes in 2016.

The FCA sandbox allows firms to test innovative products with real customers under a waiver or modification of regulatory requirements, subject to safeguards. The program has been influential globally, with sandboxes adopted in Singapore, Australia, Canada, the European Union, and many other jurisdictions. The FCA model demonstrated that it was possible to encourage innovation while maintaining meaningful consumer protections — and that regulators did not have to choose between safety and experimentation.

In the United States, the sandbox concept has been adopted primarily at the state level. Arizona was the first U.S. state to establish a regulatory sandbox, enacting legislation in 2018 that allows companies to test innovative financial products with up to 10,000 customers for up to two years without obtaining the licenses normally required to offer those products. During the sandbox period, the Arizona Attorney General's office monitors participating companies and works with them to identify the regulatory pathway they will follow if they seek to continue operations beyond the sandbox period.

Other states have followed Arizona's lead, including Utah, Wyoming, and a number of others. The state-level sandbox approach reflects the federal structure of financial regulation in the United States — because many of the most burdensome licensing requirements are at the state level, state sandboxes can provide meaningful relief. Federal sandboxes are more limited in their legal effect, because federal agencies generally cannot waive statutory requirements without Congressional authorization.

At the federal level, both the OCC and the CFPB have developed programs that function similarly to sandboxes in some respects. The OCC's responsible innovation framework and its Office of Innovation provide a mechanism for FinTech companies to engage with the OCC informally before pursuing formal licensing. The CFPB's Project Catalyst and No-Action Letter program serve a similar function in the consumer protection space. These programs do not provide formal legal immunity from regulatory requirements, but they offer a channel for dialogue that reduces uncertainty.

The value of regulatory sandboxes — and the debates about their limitations — turns on a set of practical questions. Do they actually reduce the cost and time of getting innovative products to market? Do they enable products to be tested and introduced that otherwise would never have emerged? Do they provide a forum where regulators and innovators can work together to build appropriate consumer protections into new products from the start? The evidence from the UK and from state sandbox programs in the United States suggests that the answer to all three questions is yes, at least for some companies and some products. Sandboxes are not a solution to all regulatory complexity, and they are not appropriate for every type of innovation. But they represent a meaningful shift in how regulators think about their role — from gatekeepers who approve products only after they are fully developed to collaborators who engage with innovators throughout the development process.

No-Action Letters and Informal Guidance

Regulatory uncertainty — not knowing whether a product will be treated as a security, a money transmitter service, or something else entirely — is one of the most significant barriers to FinTech innovation. Companies that cannot get a clear answer about how their product will be regulated face a difficult choice: proceed and risk enforcement action, or hold back and forfeit the market opportunity. Neither option is good.

No-Action Letters are one of the primary tools regulators use to address this problem. As noted in the discussion of the CFPB above, a No-Action Letter is a formal regulatory commitment not to recommend enforcement action against a company for a specific product or activity, provided the company operates it as described. The SEC has a long-established No-Action Letter process for securities law questions. The

CFPB formalized its process in 2016. FinCEN and other agencies have similar mechanisms, though they vary in form and legal effect.

The practical value of a No-Action Letter is significant. It allows a company to enter the market with confidence that a specific regulatory concern has been addressed. It provides documentation that the company acted in good faith, which is relevant in any subsequent enforcement proceeding. And it creates a public record that other companies in similar situations can reference — one company's No-Action Letter can provide informal guidance to the entire sector.

Informal guidance — through published FAQs, staff guidance, speeches, and industry outreach — is a less formal but in practice very important source of regulatory clarity. Regulators communicate their views on novel issues through a variety of channels short of formal rulemaking, and FinTech companies that track these communications carefully gain an interpretive advantage over those that do not.

Innovation Offices

Several federal agencies have established dedicated offices to engage with FinTech firms. The OCC's Office of Innovation is among the most prominent, serving as a central point of contact for FinTech companies seeking guidance on how OCC-supervised banks can work with them, what a special purpose charter might look like for their business, and how OCC regulations would apply to their specific product or service.

The CFPB's Project Catalyst performs a similar function on the consumer protection side, with an additional research mission — the bureau uses Project Catalyst to study how new FinTech products affect consumers, both to inform its own rulemaking and to provide guidance to companies developing those products. The CFPB has used Project Catalyst to examine innovations in credit access, savings products, and payment services, and its findings have shaped the bureau's policy positions on several important issues.

The SEC's Office of Innovation and Financial Technology provides guidance on securities law questions for FinTech firms, with particular focus on digital assets and cryptocurrency. Given the volume of enforcement actions the SEC has brought against crypto companies in recent years, the existence of a channel for proactive engagement has been important for companies that want to understand the SEC's views before making business decisions that could expose them to liability.

The OCC Special Purpose FinTech Charter

One of the most significant regulatory developments of the past decade for the FinTech industry was the OCC's 2016 proposal to offer a special purpose national bank charter to FinTech companies. The proposal represented a potential solution to one of the most fundamental problems in FinTech regulation: the fact that FinTech companies that perform banking functions do not have access to the federal banking license that would streamline their regulatory obligations and give them access to the national banking infrastructure.

What the Charter Would Offer

A special purpose national bank charter would allow a FinTech company to receive a federal banking license without needing to accept deposits. Under the OCC's proposal, a company that receives a special purpose charter would be subject to OCC supervision and would need to meet requirements in three areas: having adequate capital and liquidity; having a financial inclusion commitment; and having a recovery and resolution plan that describes how the company would wind down operations in an orderly manner if it failed.

The potential benefits are significant. A nationally chartered FinTech bank could operate in all fifty states without obtaining state money transmitter licenses or navigating state-by-state lending rate caps. It would have access to the national payment infrastructure on terms comparable to other OCC-supervised institutions. It would have a clear regulatory relationship with a single primary federal supervisor, reducing the uncertainty of operating across multiple overlapping jurisdictions.

Advantages of Being a Bank

For a FinTech company, becoming a bank offers several competitive advantages beyond regulatory streamlining. Access to government-insured deposits allows a bank to fund its operations at a lower cost than a non-bank competitor that must raise capital in the market. The ability to operate nationwide without state-by-state licensing reduces both cost and administrative burden. And the regulatory imprimatur of a bank charter can itself be a competitive asset — consumers and business partners may place greater trust in a federally supervised institution than in an unregulated FinTech startup.

These advantages come with significant costs and obligations, however. A bank charter subjects the holder to regular examination by federal regulators, capital and liquidity requirements, compliance with a broad range of consumer protection and safety-and-soundness standards, and public disclosure requirements. For a startup that values speed and flexibility above all else, the constraints of bank regulation can feel like the opposite of what FinTech is supposed to be about.

Legal and Political Obstacles

The OCC's special purpose charter proposal has faced legal and political opposition since it was announced. State banking regulators — represented by the Conference of State Bank Supervisors — have sued to block the charter on the grounds that the OCC lacks statutory authority to grant a bank charter to a company that does not accept deposits. Federal courts have reached different conclusions on this question, and the legal status of the charter remains contested as of early 2026.

The political environment around FinTech charters has also shifted. The charter proposal was developed under one administration and has received varying degrees of support from successive administrations.

The broader debate about the appropriate role of federal versus state regulation in financial services — a debate that runs through American financial history — shapes how the charter question is perceived by policymakers and industry alike.

Despite these obstacles, the special purpose charter concept remains relevant as a framework for thinking about how FinTech companies might eventually obtain the regulatory stability of a federal license without becoming full-service depository banks. Whether through the OCC charter, through ILC applications, or through future legislative action, the question of how FinTech firms can access the benefits of bank status is unlikely to go away.

Bank-FinTech Partnerships and Due Diligence

While the debate over special purpose charters continues, most FinTech companies that need access to banking infrastructure have pursued a more practical path: partnership with an existing chartered bank. These partnerships take many forms — a FinTech company might use a bank partner to hold customer deposits, to originate loans, to process payments through the ACH network, or to provide FDIC deposit insurance for customer funds. The bank provides the regulatory permissions; the FinTech provides the technology and the customer relationship.

These partnerships have become a cornerstone of the FinTech industry. But they come with a significant practical constraint: the time and cost of bank due diligence. Before a bank enters into a partnership with a FinTech company, it is required by its regulators to conduct thorough due diligence on the FinTech firm's financial condition, compliance program, technology systems, and management team. The bank's regulators hold the bank responsible for the activities of its FinTech partners, which means banks have strong incentives to be thorough — sometimes painfully thorough — in their review process.

This due diligence requirement is one of the most common factors leading to delays in bank-FinTech partnerships. A startup that expects to close a banking partnership in weeks may find the process takes months. The bank's compliance and legal teams may require extensive documentation, third-party audits, and multiple rounds of review before approving a new partner. For an early-stage company operating on venture capital runway, this delay can be consequential. And for the banking industry, the due diligence burden reflects a genuine regulatory reality — banks that partner with FinTech companies without adequate oversight have faced supervisory criticism and enforcement action when those partnerships produced consumer harm.

The U.S. Approach in International Context

The U.S. regulatory framework for FinTech is distinctive in its fragmentation, its reliance on multiple agencies with overlapping jurisdictions, and its heavy dependence on state-level regulation. Understanding how this compares to approaches in other major economies helps clarify both the challenges and the choices embedded in the U.S. system.

The United Kingdom

The United Kingdom's approach to FinTech regulation is often cited as a global benchmark for flexibility and innovation-friendliness. The Financial Conduct Authority, which has comprehensive authority over most financial services activities in the UK, developed the world's first major regulatory sandbox in 2016 and has consistently positioned itself as a regulator that wants to understand and accommodate financial innovation rather than simply restrain it.

The FCA's regulatory sandbox has been a model for programs adopted around the world. Its willingness to engage with crypto assets, open banking, and emerging payment technologies at an early stage — providing guidance and waivers that allow companies to test products before full regulatory compliance is required — has helped make London one of the world's leading FinTech hubs. The UK's single-regulator model, as contrasted with the multi-agency U.S. system, also reduces the compliance burden for firms that need clarity on a novel product: they have one regulator to engage with rather than five.

The UK's post-Brexit regulatory environment has introduced new uncertainty, as rules that previously aligned with EU frameworks have diverged in some areas. But the FCA remains a frequently cited comparison point for U.S. policymakers and industry advocates arguing for a more streamlined domestic regulatory approach.

The European Union

The European Union has taken a more prescriptive and harmonized approach to FinTech regulation than either the United States or the United Kingdom. The EU's General Data Protection Regulation, which took effect in 2018, established comprehensive data privacy rights for individuals and imposed significant compliance obligations on any company that collects or processes the personal data of EU residents — including U.S.-based FinTech companies that serve European customers. The GDPR's extraterritorial reach has made it the de facto global benchmark for data privacy, and its standards have influenced U.S. state-level privacy legislation, including the California Consumer Privacy Act.

On the payments side, the EU's Payment Services Directive — in its second iteration known as PSD2 — has established a framework for open banking that requires banks to share customer financial data with authorized third-party providers, with customer consent. This has created a more structured and legally certain open banking environment in Europe than exists in the United States, where open banking has developed largely through market dynamics and CFPB guidance rather than statutory mandate.

The EU has also adopted the Markets in Crypto-Assets regulation, known as MiCA, which provides a comprehensive regulatory framework for crypto assets across all EU member states. MiCA establishes licensing requirements for crypto asset service providers, disclosure standards for token issuers, and rules for stablecoins. Its adoption represents a level of regulatory clarity for the crypto sector that has not yet been achieved in the United States, where jurisdictional disputes between the SEC and CFTC have left significant uncertainty.

What These Comparisons Reveal

The comparison between the U.S. system and those of the UK and EU reveals several things. First, the fragmentation of U.S. financial regulation is not inevitable — other major economies have chosen more unified regulatory structures, and those structures offer some real advantages for FinTech firms seeking regulatory clarity. Second, the U.S. system's federal structure creates a layer of state-level complexity that does not have a direct parallel in most other jurisdictions. Third, and perhaps most importantly, the U.S. system's reliance on enforcement action rather than proactive licensing or sandbox engagement as the primary tool for setting regulatory boundaries creates uncertainty for innovators in a way that the UK and EU models, at their best, avoid.

None of this means the U.S. system should simply be replaced with a European or British model. The U.S. system reflects choices about federalism, separation of powers, and the structure of the financial industry that have deep roots. But the comparison is useful for understanding what alternatives exist, what trade-offs are involved, and where the U.S. approach creates unnecessary friction for legitimate innovation.

No One-Size-Fits-All Approach

A recurring theme in any serious analysis of FinTech regulation is that the appropriate regulatory treatment depends heavily on context. A FinTech company that offers a mobile payment app to consumers faces a fundamentally different regulatory situation than one that offers cryptocurrency trading, robo-advisory services, or peer-to-peer lending. Even within a single sector, the regulatory treatment of a given company depends on its specific business model, its customer base, the structure of its products, and the states in which it operates.

This variability is sometimes characterized as a problem — as evidence of a system that is too complicated, too unpredictable, and too dependent on regulatory interpretation. There is truth in that characterization. But there is also a principled rationale for a regulatory approach that calibrates requirements to risk. A company that holds customer deposits and makes loans poses different risks to consumers and to the financial system than one that merely facilitates payments between parties who maintain their own bank accounts. A one-size-fits-all approach that imposed bank-level regulation on every company that touches financial data would eliminate a large portion of the innovation that has made FinTech valuable.

The challenge is to find the right calibration — regulation that is demanding enough to prevent harm and flexible enough to accommodate legitimate innovation. That challenge does not have a single answer. It requires ongoing engagement between regulators, innovators, consumer advocates, and policymakers. It requires regulatory tools — sandboxes, No-Action Letters, innovation offices — that allow the system to evolve without waiting for the full legislative process every time a new product emerges. And it requires FinTech professionals who understand the regulatory landscape well enough to navigate it thoughtfully rather than simply trying to avoid it.

Key Takeaways

The U.S. financial regulatory landscape is defined by fragmentation: multiple federal agencies share jurisdiction over different aspects of financial services, and state governments retain significant authority over banking, lending, and money transmission. For FinTech companies, this fragmentation creates compliance complexity that is a defining feature of operating in the U.S. market.

The major federal agencies with FinTech relevance include the OCC, the Federal Reserve, the FDIC, the SEC, the CFTC, the CFPB, FinCEN, and the FTC. Each has a distinct mandate and jurisdiction. The appropriate regulatory framework for any given FinTech company depends on what it does, not simply what it calls itself.

Regulators have developed tools to engage with FinTech innovation, including regulatory sandboxes, No-Action Letters, and dedicated innovation offices. Arizona was the first U.S. state to establish a regulatory sandbox. At the federal level, the CFPB's Project Catalyst and No-Action Letter program and the OCC's Office of Innovation represent the most significant mechanisms for FinTech engagement.

The OCC's special purpose FinTech charter offers a potential pathway for FinTech companies to obtain national bank status without accepting deposits, but the proposal has faced legal and political obstacles. Bank-FinTech partnerships remain the primary practical pathway for accessing banking infrastructure, though the due diligence requirements of bank partners are a common source of delay and cost.

Compared to the UK's single-regulator model and the EU's harmonized frameworks, the U.S. system is more fragmented and more dependent on enforcement rather than proactive licensing. These differences reflect choices about federalism and regulatory structure that have real implications for the competitive environment in which U.S. FinTech firms operate.

There is no one-size-fits-all approach to FinTech regulation. The appropriate regulatory treatment depends on the nature of the activity, the risk it poses, and the context in which it occurs. Understanding this principle — and the framework of agencies and rules that gives it practical meaning — is foundational to working professionally in the FinTech industry.

Chapter 3: Cryptocurrency — Technology, Classification, and Regulation

Introduction

In October 2008, an anonymous author writing under the name Satoshi Nakamoto published a nine-page paper titled Bitcoin: A Peer-to-Peer Electronic Cash System. The paper described a method for making electronic payments directly between parties without going through a financial institution. It proposed solving a technical problem that had long frustrated designers of digital cash systems: how to prevent the same unit of digital currency from being spent twice without relying on a central authority to verify transactions. The solution was elegant — a distributed ledger maintained by a network of computers, each holding a complete record of every transaction ever made, secured through cryptographic proof.

What Nakamoto described became Bitcoin, and Bitcoin became the foundation of an asset class that has grown to include more than 20,000 distinct cryptocurrencies, spawned trillion-dollar markets at their peak, attracted some of the most ambitious technological talent in the world, and generated some of the most spectacular financial frauds in modern history. It has also generated one of the most complex and unresolved regulatory debates in the United States.

The core of that debate is a deceptively simple question: what is a cryptocurrency? The answer determines which laws apply, which agencies have jurisdiction, and what obligations fall on the companies that create, sell, and facilitate trading in these assets. Different regulators have given different answers. Courts have reached different conclusions. Congress has debated but not yet resolved the question through comprehensive legislation. In the meantime, the industry has grown enormously, consumers have been harmed in significant ways, and the regulatory framework has tried — with mixed success — to keep pace.

This chapter provides the foundational knowledge students need to understand cryptocurrency: what it is, how it works at a conceptual level, how it differs from traditional currency, and how the legal framework attempts to classify and regulate it. It covers the Howey Test — the primary legal tool used to determine whether a crypto asset is a security — and the roles of the SEC, CFTC, FinCEN, and IRS in crypto regulation. It also covers initial coin offerings and the state-level licensing requirements that apply to virtual currency exchanges. Chapter 4 builds on this foundation by examining what happens when the regulatory framework breaks down — the frauds, collapses, and enforcement actions that have defined the sector's recent history.

What Is Cryptocurrency?

To understand cryptocurrency, it helps to start with what it is not. It is not issued by a government or central bank. It is not backed by a physical commodity like gold. It does not exist as paper notes or metal coins. It has no physical form at all. A cryptocurrency is a digital asset — a record in a database — whose authenticity is verified not by a central authority but by mathematics.

More precisely, a cryptocurrency is a digital medium of exchange that uses cryptography to secure transactions, control the creation of new units, and verify the transfer of assets. The word cryptography refers to techniques for securing information through mathematical transformations that are easy to perform but practically impossible to reverse without a key. In the context of cryptocurrency, cryptography ensures that only the rightful owner of a unit of currency can authorize its transfer, and that the historical record of all transfers cannot be altered after the fact.

Fiat Currency vs. Cryptocurrency

The most important conceptual distinction in understanding cryptocurrency is the difference between fiat currency and cryptocurrency. Understanding this distinction is not just academic — it goes to the heart of why cryptocurrency exists and why it is so difficult to regulate.

Fiat currency is money that has value because a government declares it to have value and because people accept it in exchange for goods and services. The U.S. dollar is fiat currency. It is not backed by gold or any physical commodity. Its value rests on trust — trust in the U.S. government, trust in the Federal Reserve's management of the money supply, and trust in the legal system that enforces contracts denominated in dollars. The government controls the supply of fiat currency, can create more of it, and can take it out of circulation. Banks that hold and transmit fiat currency operate under comprehensive federal and state oversight.

Cryptocurrency, at least in its original conception, is designed to be the opposite. Its supply is controlled not by a government but by an algorithm. Bitcoin, for example, has a maximum supply of 21 million coins hardcoded into its protocol — no authority can create more. Its value rests not on government backing but on the willingness of buyers and sellers to exchange it, on the security of the cryptographic system, and on the network of users who maintain and validate transactions. No central bank manages its monetary policy. No government guarantees its value.

These differences have profound regulatory implications. Fiat currency enters the financial system through regulated institutions — banks, money transmitters, payment processors — and regulators can supervise those institutions. Cryptocurrency, designed specifically to bypass those intermediaries, creates enormous challenges for the oversight mechanisms that were built around them. The fact that a Bitcoin transaction can be executed between two parties anywhere in the world, without any bank or payment processor involved, without any central record-keeper, is exactly the feature that makes Bitcoin attractive to its proponents — and exactly the feature that makes it so difficult to regulate.

How Blockchain Works

The technology underlying Bitcoin and most other cryptocurrencies is the blockchain — a distributed ledger that records every transaction ever made in a public, permanent, and tamper-resistant way. Understanding blockchain at a conceptual level is important for understanding both the appeal of cryptocurrency and the regulatory challenges it creates.

A blockchain is, at its most basic, a chain of blocks. Each block contains a set of transactions — records of who sent how much to whom. Each block also contains a cryptographic hash of the previous block, which is a mathematical fingerprint that uniquely identifies it. This chaining means that if anyone tries to alter a transaction in an earlier block, the hash of that block changes, which breaks the chain for every subsequent block. Altering history would require recomputing every subsequent block faster than the rest of the network is adding new ones — a task that is computationally prohibitive on a large, well-distributed network.

The ledger is distributed rather than centralized. Instead of being stored on a single server controlled by a bank or government, copies of the blockchain are held simultaneously by thousands of computers — called nodes — spread across the internet. Every node holds a complete record of every transaction in the history of the network. When a new transaction is proposed, nodes validate it against the rules of the protocol and against the transaction history to confirm that the sender actually owns the funds being sent. Once validated, the transaction is bundled into a new block and added to the chain on every node simultaneously.

This architecture produces several properties that are central to Bitcoin's design and directly relevant to its regulatory treatment. First, Bitcoin is pseudonymous — not anonymous. Transactions are recorded publicly on the blockchain and are visible to anyone. Every transaction in the history of the Bitcoin network is visible to all nodes. However, the identities associated with Bitcoin addresses are not automatically disclosed. A Bitcoin address is a string of letters and numbers, not a name or an account number. Someone who knows only a Bitcoin address cannot immediately identify who owns it. That said, Bitcoin is not truly anonymous — blockchain analytics firms have developed sophisticated tools for tracing transaction histories and, in many cases, identifying the real-world identities behind addresses. Law enforcement has successfully used these tools in major criminal investigations.

Second, Bitcoin is decentralized. There is no central authority that controls the Bitcoin network, can freeze accounts, reverse transactions, or alter the protocol without consensus among network participants. No company owns Bitcoin. No government controls it. This decentralization is fundamental to Bitcoin's value proposition for many of its proponents — and it is also what makes traditional regulatory approaches difficult to apply directly to the currency itself.

Third, Bitcoin can be used to pay for goods and services. It is a functional medium of exchange, accepted by a growing number of merchants, used in peer-to-peer transactions, and increasingly available through payment processors that convert Bitcoin to fiat currency at the point of sale. The claim that Bitcoin cannot be used for actual commerce is simply incorrect — though it is true that price volatility makes it an impractical currency for everyday transactions for most people.

The Cryptocurrency Ecosystem

Bitcoin was the first cryptocurrency, but it is far from the only one. Today, more than 20,000 distinct cryptocurrencies exist, though the vast majority have little value or active use. The ecosystem has grown from a single currency into a complex landscape of competing projects with different technical architectures, different purposes, and different regulatory profiles.

Ethereum, launched in 2015, introduced a significant innovation: the smart contract. A smart contract is a self-executing program stored on the blockchain that automatically carries out the terms of an agreement when specified conditions are met. Smart contracts make it possible to build decentralized applications — financial products, markets, and services — that run on the blockchain without any central operator. Ethereum's smart contract platform became the foundation for the initial coin offering boom of 2017 and for the decentralized finance ecosystem that has since emerged.

Stablecoins are a category of cryptocurrency designed to maintain a stable value by pegging to a reference asset — typically the U.S. dollar. Tether and USD Coin are among the most widely used. Stablecoins serve an important function within crypto markets: they allow traders and investors to hold value in a relatively stable form without converting back to fiat currency. They also facilitate cross-border payments and have attracted significant regulatory attention because of the systemic risks they could create if their dollar pegs were to break.

Non-fungible tokens, or NFTs, are digital assets recorded on a blockchain that represent ownership of a unique item — a piece of digital art, a collectible, a piece of virtual real estate. Unlike Bitcoin or Ethereum, which are fungible (one unit is interchangeable with any other), each NFT is unique. NFTs attracted enormous speculative interest between 2020 and 2022 before experiencing a significant market decline.

Decentralized finance, or DeFi, refers to a broad category of financial services — lending, borrowing, trading, earning interest — that are conducted through smart contracts on a blockchain, without any centralized intermediary. DeFi platforms do not have corporate headquarters, customer service departments, or compliance officers.

They exist as code on a blockchain. This architecture creates profound regulatory challenges, because the enforcement tools regulators typically use — licensing requirements, examination authority, enforcement against corporate entities — have limited applicability to a protocol that has no identifiable operator.

This diversity matters for regulation. A cryptocurrency that functions primarily as a medium of exchange occupies a different legal space than one that represents ownership of a profit-generating enterprise. A stablecoin that is backed by dollar reserves raises different concerns than one backed by an algorithmic mechanism. A DeFi lending protocol raises different questions than a centralized exchange with identifiable owners and operators. Understanding the ecosystem means recognizing that cryptocurrency is not a monolithic category and that regulatory analysis must be tailored to the specific characteristics of each asset and platform.

The Classification Problem

The most consequential regulatory question in cryptocurrency is also the most basic: what is it? The answer is not obvious, and the various possible answers lead to very different regulatory outcomes. Is a cryptocurrency a currency? A commodity? A security? Property? The answer determines which agency has jurisdiction, what disclosure and registration requirements apply, what consumer protections are available, and how the asset is taxed.

The difficulty is that cryptocurrency does not fit cleanly into any of the categories that existing law was designed to address. It shares some features with currency — it can be used as a medium of exchange — but it is not issued by a sovereign government and is not legal tender. It shares some features with commodities — it can be traded on exchanges, its price fluctuates based on supply and demand, and futures contracts can be written on it — but it is not a physical good. It shares some features with securities — many crypto projects are sold to investors with expectations of profit — but not all of them, and the legal test for whether something is a security was developed long before digital assets existed.

Global regulators have not reached a consensus on how to classify cryptocurrency. Different countries have made different choices, and within the United States, different federal agencies have staked out positions that are not always consistent with each other. This lack of consensus is itself a significant regulatory reality — it means that a U.S. company operating in the crypto space may face conflicting regulatory signals from multiple agencies simultaneously.

The table below summarizes the major classification options and their regulatory implications.

Classification	Primary Regulator	Legal Framework	Key Implication
Security	SEC	Securities Act of 1933; Securities Exchange Act of 1934	Must register offering or qualify for exemption; platforms trading it must register as exchanges or broker-dealers
Commodity	CFTC	Commodity Exchange Act	Spot markets largely unregulated federally; derivatives markets subject to CFTC oversight
Currency / Money Transmitter	FinCEN; state regulators	Bank Secrecy Act; state money transmitter laws	Must register with FinCEN as money services business; obtain state money transmitter licenses
Property	IRS	Internal Revenue Code	Subject to capital gains taxation; gains and losses must be reported

The Securities Question and the Howey Test

The most contested classification question in U.S. crypto regulation is whether a given cryptocurrency is a security. If it is, it falls under the jurisdiction of the Securities and Exchange Commission, or SEC, and must comply with the registration and disclosure requirements of the federal securities laws — requirements that are comprehensive, expensive, and largely incompatible with the way most crypto assets have been sold to date.

The legal test for whether an asset is a security comes from a 1946 Supreme Court case called SEC v. W.J. Howey Co. The Howey case involved the sale of plots of a Florida citrus grove combined with a service contract under which the seller would cultivate and harvest the oranges and share the profits with buyers. The Supreme Court held that this arrangement was an investment contract — a form of security — and articulated a four-part test for identifying investment contracts that has been applied to novel financial instruments ever since.

Under the Howey Test, an arrangement constitutes an investment contract — and therefore a security — if it involves: (1) an investment of money; (2) in a common enterprise; (3) with an expectation of profit; (4) from the efforts of others. All four elements must be present. If any one is absent, the instrument is not an investment contract under Howey.

The table below breaks down each element and its application to cryptocurrency.

Component	What It Requires	Application to Crypto
Investment of money	A person must contribute money or other value	Satisfied when investors purchase tokens with fiat currency or other cryptocurrency
In a common enterprise	The investor's fortunes must be tied to the fortunes of others in the enterprise	Satisfied when token value depends on the pooled efforts of developers, promoters, or a central team

With an expectation of profit	The investor must expect to earn a return on the investment	Satisfied when tokens are marketed as appreciating in value or generating returns
From the efforts of others	The expected profit must derive primarily from the efforts of a third party, not the investor	This is the most contested element in crypto; satisfied when investors are passive and depend on developers or promoters to create value

The fourth element — profit from the efforts of others — is typically the most contested in crypto cases. The SEC has argued that most cryptocurrencies sold in initial coin offerings satisfy this element because buyers are passive investors whose returns depend entirely on the work of the development team. Token issuers have argued the opposite — that once a network is sufficiently decentralized, no identifiable group is responsible for its value, and the efforts-of-others element is not met.

The SEC's former chairman William Hinman offered a notable articulation of this view in a 2018 speech, suggesting that a cryptocurrency that was initially a security might cease to be one if the network becomes sufficiently decentralized over time — because at that point, no central promoter's efforts drive the value of the asset. The speech was not formal guidance, but it has been widely cited in litigation and policy debates. The SEC has since distanced itself from the Hinman framework to some extent, but the underlying question — at what point, if ever, does decentralization remove a crypto asset from securities regulation — remains unresolved.

It is critical to note what the Howey Test does not say. The fourth component is profit from the efforts of others — not, as a common distractor might suggest, governance by a board of directors. Corporate governance structures are irrelevant to the Howey analysis. What matters is whether the investor's expected return depends on the entrepreneurial or managerial efforts of someone other than the investor.

The Commodity Classification

The Commodity Futures Trading Commission, or CFTC, has taken the position that Bitcoin and Ethereum are commodities under the Commodity Exchange Act. This determination rests on the broad definition of commodity in federal law, which includes a wide range of goods and financial instruments, and on the fact that these cryptocurrencies function as stores of value and media of exchange rather than as investment contracts in the Howey sense.

The CFTC's jurisdiction over cryptocurrency is primarily activated through derivatives. The CFTC has clear authority over futures contracts, options, and swaps written on any commodity — including crypto commodities. Bitcoin futures have been traded on CFTC-regulated exchanges since 2017. A FinTech company that offers cryptocurrency derivatives products to U.S. customers must register with the CFTC as a designated contract market, swap execution facility, or futures commission merchant, depending on the structure of its operations.

The CFTC's authority over spot markets — markets for the immediate purchase and sale of cryptocurrency — is more limited. The CFTC can bring fraud and manipulation cases in spot markets, but it does not have the same comprehensive registration and oversight authority over spot trading that the SEC has over securities markets. This gap has been a persistent source of regulatory concern, because most retail crypto trading occurs in spot markets rather than derivatives markets.

Currency and Money Transmission

Even if a cryptocurrency is not a security or a commodity in the traditional sense, the act of transmitting it — moving it from one party to another on behalf of customers — implicates the money transmission regulatory framework. The Financial Crimes Enforcement Network, or FinCEN, has made clear that virtual currency administrators and exchangers are money services businesses under the Bank Secrecy Act and must register with FinCEN, implement anti-money laundering programs, and file suspicious activity reports.

A virtual currency user — someone who obtains and uses cryptocurrency for personal transactions — is not a money services business and is not subject to Bank Secrecy Act registration requirements. But a virtual currency exchanger — a business that exchanges cryptocurrency for fiat currency or for other cryptocurrency on behalf of customers — and a virtual currency administrator — a business that issues and redeems a centralized virtual currency — are both classified as money services businesses with full Bank Secrecy Act obligations.

This distinction matters because it defines the scope of FinCEN's reach into the crypto ecosystem. Centralized exchanges — platforms like Coinbase that hold customer funds and facilitate trading — are clearly money services businesses. Decentralized exchanges that operate through smart contracts without any central operator present a more difficult question, and FinCEN and other regulators have been working to develop guidance on how the money transmission framework applies to DeFi.

Cryptocurrency as Property

The Internal Revenue Service, or IRS, has taken the position that cryptocurrency is property for federal tax purposes. This determination, first announced in IRS Notice 2014-21, has significant practical implications for anyone who buys, sells, or uses cryptocurrency.

Because cryptocurrency is property rather than currency, every transaction involving cryptocurrency is a taxable event. If you buy Bitcoin at $20,000 and later sell it at $60,000, you have a $40,000 capital gain that must be reported on your federal tax return. If you use Bitcoin to pay for goods or services, the difference between the value of the Bitcoin when you acquired it and its value when you spent it is a realized gain or loss. Even exchanging one cryptocurrency for another — say, trading Bitcoin for Ethereum — is a taxable event.

This treatment creates significant record-keeping burdens for active crypto users and traders, and it stands in contrast to the treatment of foreign currency, gains on which are not taxed in most consumer transactions below a certain threshold. The IRS has increasingly focused on crypto tax compliance, requiring exchanges to issue tax forms to customers and requiring taxpayers to disclose their cryptocurrency activity on their federal returns.

The Role of Federal Regulators

Given the classification complexity described above, it is not surprising that multiple federal agencies claim some form of jurisdiction over cryptocurrency. Understanding which agency does what — and where jurisdictional gaps and overlaps exist — is essential for any professional working in or with the crypto sector.

The SEC's Role

The SEC's position is that most cryptocurrencies sold to investors — particularly through initial coin offerings — are securities, and that the companies that sold them were required to register those offerings or qualify for an exemption. The SEC has pursued this position aggressively through enforcement. It has brought actions against dozens of crypto companies for selling unregistered securities, against exchanges for operating unregistered securities markets, and against promoters for market manipulation and fraud.

The SEC's enforcement posture shifted notably after 2024. Under new leadership, the agency began withdrawing some pending enforcement actions against crypto companies and signaling a more accommodating approach to the industry. The broader direction of SEC crypto policy as of early 2026 reflects ongoing political and regulatory recalibration rather than settled doctrine. What has not changed is the underlying legal framework: the Howey Test remains the standard for determining whether a crypto asset is a security, and the consequences of selling unregistered securities remain serious.

The SEC's approach to crypto also extends to trading platforms. An exchange that lists securities — including crypto assets that qualify as securities — must register as a national securities exchange or operate under an exemption. The SEC has argued that many crypto exchanges are operating as unregistered securities exchanges, creating significant legal exposure for those platforms.

The CFTC's Role

The CFTC's jurisdiction over crypto is narrower but well-established in the derivatives context. Bitcoin futures, Ethereum futures, and other crypto derivatives traded on U.S. markets are subject to CFTC oversight. The CFTC has also brought enforcement actions in spot markets for fraud and manipulation, asserting its anti-fraud authority even where its full registration and oversight authority does not apply.

The jurisdictional boundary between the SEC and CFTC in crypto has been a source of persistent friction. Both agencies have claimed authority over certain assets, and they have not always reached consistent conclusions. Congress has debated legislation — including the CLARITY Act, discussed in Chapter 4 — that would clarify the jurisdictional boundary by establishing a clear framework for determining when a crypto asset is a commodity subject to CFTC oversight versus a security subject to SEC oversight.

FinCEN's Role

FinCEN's primary role in crypto regulation is through the Bank Secrecy Act's anti-money laundering framework. Every virtual currency exchange operating in the United States must register with FinCEN as a money services business, implement a written anti-money laundering program, conduct customer due diligence, file currency transaction reports for transactions above $10,000, and file suspicious activity reports when transactions suggest possible illegal activity.

FinCEN's reach is significant because it applies regardless of how a crypto asset is classified under securities or commodities law. A cryptocurrency exchange that is not selling securities — and therefore not subject to SEC jurisdiction — is still operating as a money services business and must comply with FinCEN requirements. This means that virtually every centralized crypto exchange operating in the United States has mandatory FinCEN obligations, regardless of the securities status of the assets it trades.

FinCEN has proposed and in some cases finalized rules extending Bank Secrecy Act requirements to new areas of the crypto ecosystem, including rules requiring exchanges to collect and transmit identifying information about the parties to certain transactions. These so-called travel rule requirements — already applicable to wire transfers in traditional banking — have been extended to virtual currency transfers, creating compliance challenges for exchanges that interact with unhosted wallets, where the recipient is not a customer of any regulated institution.

The IRS's Role

The IRS enforces the tax treatment of cryptocurrency as property. Its practical tools include information reporting requirements — exchanges must issue Form 1099 to customers who sell cryptocurrency — and audit and enforcement authority against taxpayers who fail to report crypto gains. The IRS has also used John Doe summonses to compel exchanges to produce customer records, allowing the agency to identify taxpayers who received cryptocurrency but failed to report it.

For FinTech companies that facilitate cryptocurrency transactions, IRS compliance obligations include both their own tax treatment and their information reporting obligations to customers and to the IRS. Companies that treat crypto payments as business income, that hold crypto on their balance sheets, or that offer crypto-related financial products face complex accounting and tax questions that are still being worked out in guidance and litigation.

State Regulation of Virtual Currency

Federal regulation of cryptocurrency is only part of the picture. States have their own regulatory frameworks for financial services, and most states require money transmitters — including virtual currency exchanges — to obtain state licenses before operating. Navigating this state-level landscape is one of the most operationally demanding aspects of running a virtual currency business in the United States.

State Money Transmitter Licensing

As discussed in Chapter 2, most states require money transmitters to obtain a license before conducting money transmission business within the state. For virtual currency exchanges, this requirement creates a significant compliance burden: a company that wants to serve customers in all fifty states must obtain and maintain money transmitter licenses — or their equivalent — in each state that requires them.

The requirements vary considerably. Some states have relatively streamlined processes and modest financial requirements. Others impose substantial net worth requirements, surety bond obligations, and examination processes that can take months. The inconsistency across states is a persistent source of frustration for the industry and a significant barrier to entry for smaller companies that cannot afford the compliance infrastructure required to operate nationally.

Not all states have clarified whether their existing money transmitter statutes apply to virtual currency. Some have enacted specific legislation addressing virtual currency. Others have issued guidance interpreting existing statutes to cover it. A few have taken the position that their existing laws do not apply. For a virtual currency exchange seeking to operate in every U.S. state, the regulatory status in each state must be determined individually — a process that typically requires legal analysis and sometimes direct engagement with state regulators.

The New York BitLicense

New York's approach to virtual currency regulation stands apart from that of every other state. In 2015, the New York State Department of Financial Services introduced the BitLicense — a specific regulatory framework for virtual currency businesses that operate in New York or serve New York customers. The BitLicense requires virtual currency businesses to obtain a specific license from the Department of Financial Services, meet capital and liquidity requirements, implement anti-money laundering and customer identification programs, maintain detailed records, submit to regular examination, and comply with consumer protection requirements.

The BitLicense was the first comprehensive state-level regulatory framework specifically designed for virtual currency, and it generated significant controversy when it was introduced. Critics argued that its requirements were so burdensome that they would drive crypto businesses out of New York — and indeed, several prominent companies announced they would not seek a BitLicense and would instead stop serving New York customers. Proponents argued that it provided necessary consumer protections and gave compliant businesses a regulatory seal of approval.

The BitLicense has since become a benchmark — cited both as a model of proactive state regulation and as a cautionary tale about the potential for state-level requirements to fragment the national market. For any virtual currency exchange that wants to serve customers in New York — one of the largest financial markets in the world — obtaining the BitLicense is not optional. As of early 2026, only a limited number of companies have received BitLicenses, reflecting both the rigor of the application process and the ongoing consolidation of the regulated crypto exchange market.

The Broader State Landscape

Beyond money transmission licensing and the New York BitLicense, states have engaged with cryptocurrency in a variety of ways. Some states — Wyoming most prominently — have enacted legislation specifically designed to attract crypto and blockchain businesses, creating favorable frameworks for digital asset custody, crypto-friendly bank charters, and decentralized autonomous organization structures. Others have been more cautious or restrictive. The result is a patchwork of state-level approaches that adds to the complexity already created by federal regulatory fragmentation.

State securities regulators have also been active in the crypto space, bringing enforcement actions against initial coin offerings and other crypto investment products that they view as unregistered securities under state law. The North American Securities Administrators Association, which represents state securities regulators, has coordinated multi-state enforcement actions against crypto fraud, demonstrating that state-level securities regulation is a live enforcement risk for crypto companies — not just a theoretical one.

Initial Coin Offerings

One of the most consequential developments in the history of cryptocurrency regulation was the initial coin offering boom of 2016 and 2017. Understanding what an ICO is, how the regulatory framework applies to it, and what went wrong during this period is essential background for the case studies in Chapter 4.

What Is an Initial Coin Offering?

An initial coin offering, or ICO, is a fundraising mechanism in which a company or project sells newly created cryptocurrency tokens to investors, typically in exchange for established cryptocurrencies like Bitcoin or Ethereum. The name is a deliberate echo of the initial public offering, or IPO, through which companies sell stock to the public — but the legal structure is entirely different.

In a typical ICO, a project team publishes a whitepaper describing the technology they plan to build, the problem it will solve, and the role that the token will play in the resulting ecosystem. Investors who believe in the project purchase tokens during the offering. The team uses the proceeds to fund development. If the project succeeds and the token becomes widely used, its value increases — generating returns for early investors. If the project fails, the tokens may become worthless.

ICOs proved enormously popular during 2017. The ease of creating tokens on the Ethereum blockchain, the global accessibility of the offering mechanism, and the speculative frenzy of the crypto bull market combined to produce a fundraising wave in which thousands of projects raised billions of dollars from retail investors around the world — largely without any regulatory oversight.

Pre-Release and Post-Release ICOs

ICO structures vary in important ways, and the distinction between pre-release and post-release offerings has both practical and regulatory significance.

A pre-release ICO — sometimes called a pre-sale or a sale of development-stage tokens — occurs before the underlying platform or protocol has been built. Investors are essentially funding the creation of a product that does not yet exist. The tokens they receive have no immediate utility; their value is entirely speculative, based on the expectation that the development team will successfully build the promised platform and that the resulting tokens will have value within it. This is the structure that most closely resembles an investment contract under the Howey Test, because investors are clearly depending on the efforts of the development team to create value.

A post-release ICO occurs after the platform has been built and the tokens have immediate utility within a functioning ecosystem. Buyers can use the tokens immediately for the purpose the platform is designed for, rather than waiting for a development team to build something. The SEC has suggested that post-release tokens — particularly those on sufficiently decentralized networks — may be less likely to qualify as securities, because the investor is not solely dependent on the efforts of a central team.

In practice, the line between pre-release and post-release is often blurry. Many ICOs sold tokens that had nominal utility in a rudimentary platform while the real product was still under development. Regulators have looked to substance rather than form — to whether investors were actually buying tokens for their utility or for their investment potential — in making securities determinations.

The Regulatory Response to ICOs

The SEC moved against the ICO market in 2017 with a landmark report on the DAO — a decentralized autonomous organization that had conducted a token sale in 2016 before its funds were stolen in a hack. The SEC's DAO Report concluded that the DAO tokens were securities under the Howey Test and that the sale had been an unregistered securities offering. The report put the crypto market on notice that the SEC viewed most ICO tokens as securities and that the existing securities laws applied to their issuance and trading.

The SEC followed the DAO Report with a wave of enforcement actions against ICO issuers. Companies that had raised hundreds of millions of dollars through token sales were required to return funds to investors, pay penalties, and in some cases register their tokens as securities retroactively. The enforcement campaign effectively ended the ICO boom as a fundraising mechanism for U.S. companies and significantly chilled the market for unregistered token sales globally.

A company whose ICO is found to be an unregistered securities offering faces serious consequences: disgorgement of the funds raised, civil penalties, and potential criminal liability for executives. Investors who purchased tokens in an unregistered offering may have the right to rescind their investment and receive their money back — a potentially devastating liability for a company that has already spent the ICO proceeds on development.

ICOs have largely been replaced by alternative fundraising structures — private token sales to accredited investors, which can qualify for exemptions from securities registration requirements, and simple agreements for future tokens, which defer the token issuance until after a project is built and its regulatory status is clearer. These structures are more expensive and less accessible to retail investors, but they are more defensible under the securities laws.

Krustyburger and the Howey Test Applied

The Howey Test is best understood through application. Consider a hypothetical startup — call it Krustyburger — that offers a blockchain-based food review service and plans to raise $10 million through a token sale. Token purchasers will be able to use the tokens for in-app purchases, earn tokens by contributing reviews, and buy and sell goods and services in the platform's ecosystem. The company also tells investors that advertising revenue will increase the value of the tokens based on the platform's profitability.

Does the Krustyburger token sale constitute an investment contract? Working through the Howey elements: investors are investing money — yes. In a common enterprise — yes, the success of the platform ties investor returns together. With an expectation of profit — yes, the company has explicitly told investors that profitability will increase token value. From the efforts of others — yes, the value of the tokens depends on Krustyburger's development team building and monetizing the platform.

The fact that the tokens can be used for in-app purchases does not save the offering from securities classification if the primary reason investors are buying them is for investment returns rather than for immediate utility. The SEC has consistently held that marketing tokens as appreciating in value based on the efforts of a development team makes them securities, regardless of any incidental utility they might have. Krustyburger's claim that it need not register the offering because it is selling app-based tokens rather than shares in the company is incorrect — the Howey Test does not require an instrument to look like a stock in order to be a security.

The Path Toward Regulatory Clarity

One of the most consistent complaints from the cryptocurrency industry has been the lack of regulatory clarity — the inability to get definitive answers about how the securities laws apply to specific tokens, what registration requirements apply to specific types of platforms, and how existing rules should be interpreted in the context of genuinely novel technology. This complaint is not without merit. The SEC's primary enforcement tool in the crypto space has been enforcement action against completed transactions rather than proactive guidance about what is permitted.

Congress has introduced legislation aimed at providing the clarity the industry seeks. The CLARITY Act — introduced and debated in various forms in recent years — attempts to establish a statutory framework for classifying digital assets. Under proposals along these lines, a digital asset that functions primarily as a commodity would fall under CFTC jurisdiction, while one that functions as a security would fall under SEC jurisdiction. The framework would also create a pathway for crypto networks to transition from security classification to commodity classification as they become more decentralized over time — addressing the question that the SEC's informal Hinman framework raised but never formally resolved.

As of early 2026, comprehensive federal crypto legislation has not been enacted. The jurisdictional dispute between the SEC and CFTC, political disagreements about the appropriate level of crypto regulation, and the lobbying power of both the traditional financial industry and the crypto industry have all contributed to legislative inaction. The result is a regulatory environment defined as much by enforcement history and agency guidance as by statute — which means it can change significantly with shifts in agency leadership and administration priorities.

For FinTech professionals, the practical implication is that cryptocurrency regulation is a moving target. The foundational legal tools — the Howey Test, the Bank Secrecy Act, state money transmitter licensing — are stable. But the specific conclusions drawn from those tools, the vigor with which different agencies enforce them, and the emerging legislative frameworks that may eventually resolve current ambiguities are all subject to ongoing change. Staying current with regulatory developments is not optional in this sector — it is a core professional competency.

Key Takeaways

Cryptocurrency is a digital asset whose authenticity is verified through cryptography rather than by a central authority. It differs from fiat currency in that it is not issued by a government, is not backed by a physical commodity, and — in its original design — operates without any central institution controlling its supply or validating transactions.

Blockchain technology, the foundation of most cryptocurrencies, is a distributed ledger that records every transaction in a public, permanent, and tamper-resistant way. Bitcoin is pseudonymous rather than anonymous — transactions are publicly visible, but addresses are not automatically tied to real-world identities. Bitcoin is decentralized, meaning no central authority controls it, and it can be used to pay for goods and services.

More than 20,000 cryptocurrencies exist, ranging from Bitcoin and Ethereum to stablecoins, NFTs, and the products of decentralized finance platforms. This diversity means that regulatory analysis must be tailored to the specific characteristics of each asset — there is no uniform answer to how cryptocurrency is regulated.

The classification of a cryptocurrency determines its regulatory treatment. The Howey Test — drawn from a 1946 Supreme Court case — is the primary legal tool for determining whether a crypto asset is a security. The four elements are: an investment of money, in a common enterprise, with an expectation of profit, from the efforts of others. All four must be present. The SEC, CFTC, FinCEN, and IRS each have jurisdiction over different aspects of cryptocurrency, and their authority may overlap or conflict depending on how a particular asset is classified.

Initial coin offerings are fundraising mechanisms in which tokens are sold to investors. Pre-release ICOs — in which tokens are sold before the underlying platform exists — most closely resemble investment contracts under the Howey Test. The SEC's enforcement actions against ICO issuers effectively ended the unregistered ICO as a practical fundraising tool for U.S. companies.

At the state level, virtual currency exchanges must comply with money transmitter licensing requirements in each state where they operate. New York's BitLicense is the most comprehensive state-level framework for virtual currency businesses. Not all states have resolved how their existing money transmitter statutes apply to virtual currency, creating a patchwork that requires jurisdiction-by-jurisdiction analysis.

Global regulators disagree on how to classify cryptocurrency, and within the United States, the jurisdictional boundary between the SEC and CFTC remains contested. Comprehensive federal legislation to resolve these questions has been debated but not enacted as of early 2026. For professionals in this sector, regulatory fluency requires not just understanding the current rules but tracking the ongoing evolution of policy, enforcement, and legislation.

Chapter 4: Crypto in Practice — Fraud, Collapse, and Regulatory Response

Introduction

Chapter 3 built the legal and technical foundation for understanding cryptocurrency: how it works, how it is classified, and which agencies have jurisdiction over different types of crypto assets and activities. This chapter applies that foundation to what has actually happened — the frauds, the collapses, the enforcement actions, and the ongoing effort to build a regulatory framework capable of governing one of the most volatile and consequential sectors in modern finance.

The history of cryptocurrency is not just a story of technological innovation. It is also a story of governance failure, regulatory arbitrage, and the consequences of allowing a multi-trillion-dollar industry to grow with minimal oversight. The collapse of FTX in 2022, the SEC's protracted legal battle with Ripple over the classification of XRP, the global enforcement actions against Binance, and the failure of Celsius Network and other crypto lenders all illustrate what happens when the tools introduced in Chapter 3 — the Howey Test, AML and KYC requirements, securities registration, and exchange licensing — are either absent, ignored, or unenforced.

Understanding these cases is not merely historical. The patterns of failure described here recur across the industry. Learning to recognize them, apply the relevant regulatory frameworks, and propose better governance is a core analytical skill for anyone working in or around FinTech.

This chapter also examines where crypto regulation stands today and where it is heading. The post-2024 shift in SEC enforcement posture, the ongoing legislative debate over the CLARITY Act, and the emergence of stablecoin-specific legislation all represent meaningful changes from the enforcement-first approach that characterized the previous several years. Whether those changes represent a more rational regulatory framework or a retreat from necessary consumer protection is a question reasonable people disagree about — and one that this chapter presents in terms that allow you to form your own view.

AML and KYC Requirements for Crypto Exchanges

Before examining the major collapses, it is worth establishing in more detail how anti-money laundering and know-your-customer requirements apply to cryptocurrency exchanges. These obligations are not peripheral compliance details — they are central to how crypto platforms are regulated, and their absence or evasion is a thread running through nearly every major crypto fraud and collapse.

Why Crypto Presents Particular AML and KYC Challenges

Traditional financial institutions have operated under Bank Secrecy Act anti-money laundering obligations since 1970. Banks must verify the identity of their customers, monitor transactions for suspicious activity, file suspicious activity reports with the Financial Crimes Enforcement Network, or FinCEN, and maintain records that allow law enforcement to reconstruct the flow of funds. These requirements exist because financial institutions are the gateway through which criminal proceeds enter the legitimate financial system. If you want to launder money, you eventually need to move it through a bank.

Cryptocurrency was explicitly designed to enable peer-to-peer transactions without financial intermediaries — which means, in theory, without the oversight that those intermediaries provide. A Bitcoin transaction between two parties anywhere in the world can be executed without a bank, without identity verification, and without any report being filed with any government agency. The pseudonymous nature of blockchain addresses — discussed in Chapter 3 — means that the parties to a transaction are not automatically identifiable from the transaction record alone.

These features make cryptocurrency attractive to criminals. Drug trafficking proceeds have been laundered through crypto exchanges. Ransomware attackers demand payment in Bitcoin or privacy-focused cryptocurrencies specifically because the transactions are harder to trace than wire transfers. Sanctions evaders have used crypto to move money across borders in ways that bypass the correspondent banking system. North Korea's state-sponsored hackers have stolen billions of dollars in cryptocurrency from exchanges around the world, using the proceeds to fund weapons programs. These are not hypothetical concerns — they are documented patterns that regulators have documented extensively.

The response from U.S. regulators has been to apply the Bank Secrecy Act framework to cryptocurrency businesses as aggressively as possible within the existing legal structure. As established in Chapter 3, virtual currency exchangers and administrators are classified as money services businesses under FinCEN's rules, which means they are subject to the full range of Bank Secrecy Act obligations. What that means in practice deserves close examination.

Customer Identification and Due Diligence

Know Your Customer, or KYC, is the collective term for the identity verification and due diligence obligations that financial institutions — including crypto exchanges — must fulfill before and during customer relationships. At its most basic level, KYC means verifying that a customer is who they say they are. For a bank, this typically involves collecting government-issued identification, verifying the customer's address, and screening the customer's name against sanctions lists and databases of known criminals.

For a crypto exchange, the practical implementation of KYC creates significant friction with the pseudonymous, borderless character of cryptocurrency that many users prize. An exchange that requires customers to submit government ID before trading is no longer fully permissionless — it has become a regulated financial intermediary with obligations to know its customers and report suspicious behavior. Customers who value privacy may take their business to exchanges that impose less stringent requirements, which creates a competitive pressure against rigorous KYC compliance.

This tension is not merely theoretical. It explains why some of the largest crypto enforcement actions in history have involved KYC failures at major exchanges. When an exchange fails to verify customer identities, it cannot know whether it is facilitating transactions for drug traffickers, sanctions evaders, or terrorist financiers. And when it facilitates those transactions, it can be held liable under the Bank Secrecy Act, the International Emergency Economic Powers Act, and a range of other federal laws.

Beyond basic identity verification, effective KYC requires customer due diligence — understanding the nature and purpose of a customer relationship and monitoring it for activity inconsistent with what the exchange knows about the customer. A customer who opens an account claiming to use it for occasional personal trading but immediately begins moving large volumes of funds in patterns associated with money laundering should trigger enhanced review. This kind of ongoing monitoring requires investment in compliance infrastructure: trained personnel, transaction monitoring systems, and processes for escalating suspicious activity to senior compliance staff and, when warranted, to FinCEN.

Suspicious Activity Reporting

When a money services business — including a crypto exchange — identifies a transaction or pattern of transactions that it knows, suspects, or has reason to suspect involves funds from illegal activity, is designed to evade reporting requirements, or lacks a lawful purpose, it is required to file a Suspicious Activity Report, or SAR, with FinCEN. SARs are confidential — the customer being reported is not notified that a report has been filed — and they are the primary mechanism through which financial institutions contribute to law enforcement investigations of financial crime.

The suspicious activity reporting obligation is demanding in practice. An exchange that processes millions of transactions daily cannot have human reviewers examine each one. Compliance programs rely heavily on automated transaction monitoring systems that flag activity meeting certain risk criteria — transactions above threshold amounts, rapid movement of funds through multiple wallets, patterns associated with known money laundering typologies, and transactions involving addresses that blockchain analytics firms have flagged as associated with illicit activity.

The quality of SAR filings matters as much as the quantity. A SAR that provides only minimal information about a suspicious transaction is far less useful to law enforcement than one that provides a detailed account of the activity, the customer's history, and the basis for suspecting illegal conduct. Regulators evaluate the quality of an exchange's SAR filings as part of their assessment of its overall AML program.

The Travel Rule Applied to Crypto

One of the most operationally challenging AML requirements for crypto exchanges is the travel rule — the requirement, under the Bank Secrecy Act, that financial institutions transmitting funds above a certain threshold must pass along identifying information about the originator and beneficiary of the transfer to the receiving institution. The travel rule was designed for wire transfers, where both the sending and receiving institution are regulated entities with established compliance programs.

Applying the travel rule to cryptocurrency is technically complex because transactions often involve unhosted wallets — wallets that are not held by any regulated exchange or custodian, but rather directly controlled by the user on their own device. When a customer sends cryptocurrency from a regulated exchange to an unhosted wallet, the travel rule requires the exchange to collect and transmit information about the transaction — but there is no receiving institution on the other end to receive that information. Regulators have grappled with how to implement the travel rule in the crypto context in a way that achieves its anti-money laundering objectives without effectively prohibiting the use of unhosted wallets altogether.

FinCEN has proposed rules specifically addressing crypto travel rule compliance, and international regulatory bodies have issued guidance through the Financial Action Task Force, which sets global AML standards. As of early 2026, the implementation of travel rule requirements for crypto remains an active area of regulatory development, with exchanges investing significantly in the technical infrastructure required to comply.

The FTX Collapse: Governance Failure at Scale

No event in the history of cryptocurrency has done more to shape the current regulatory conversation than the collapse of FTX in November 2022. FTX was, at its peak, one of the largest cryptocurrency exchanges in the world — valued at $32 billion, backed by major institutional investors, and led by Sam Bankman-Fried, a figure who had cultivated a public image as a responsible actor in an industry not known for responsibility. The speed and scale of the collapse, and the nature of the misconduct that caused it, made it impossible for policymakers to continue treating crypto regulation as a deferred priority.

What FTX Was and How It Operated

FTX was a centralized cryptocurrency exchange — a platform where customers could deposit funds, trade a wide range of crypto assets, and withdraw their balances. It offered a broad array of products, including spot trading, derivatives, leveraged tokens, and staking services. At its peak, it processed tens of billions of dollars in trading volume daily and had millions of customers around the world.

FTX was incorporated in the Bahamas, a deliberate choice that allowed it to avoid the regulatory requirements that would have applied had it been headquartered in the United States. U.S. customers were technically prohibited from using FTX's primary platform and were directed to a separate U.S.-registered affiliate, FTX US, which operated under lighter regulatory constraints than a fully licensed U.S. exchange would have faced. This corporate structure — headquartered offshore, with a nominally separate U.S. entity — was a form of regulatory arbitrage that allowed the company to serve global markets while minimizing its regulatory footprint in the most demanding jurisdictions.

FTX was closely affiliated with Alameda Research, a quantitative trading firm also controlled by Bankman-Fried. The relationship between the two entities was central to what ultimately destroyed both of them.

The Misconduct: Customer Funds and the Alameda Relationship

The core of the FTX fraud was straightforward, even if it was hidden behind layers of complexity: customer funds deposited at FTX were being transferred to Alameda Research and used for Alameda's trading activities, investments, and operational expenses — without customer knowledge or consent. Customers believed their funds were being held in custody by FTX, available for trading or withdrawal at any time. In reality, a substantial portion of those funds had been lent to Alameda, which had used them in ways that left FTX unable to meet customer withdrawal requests when they came due.

The mechanism that enabled this was a secret back-door access that Alameda had to FTX's systems — access that allowed Alameda to draw on customer funds held at the exchange without going through normal processes that would have made the transfers visible to outside auditors or compliance personnel. FTX's internal controls were, by virtually any measure, nonexistent. There was no independent board of directors with meaningful oversight authority. There was no chief financial officer with authority to question how funds were being used. The company's financial accounts were maintained in informal ways that made it effectively impossible to determine, from the outside, how much money was where.

When a journalist published an analysis in November 2022 suggesting that Alameda's balance sheet was dominated by FTT — a token issued by FTX itself — the market recognized immediately that both entities were far more financially precarious than they appeared. Customers attempted to withdraw their funds en masse. FTX could not meet the withdrawal requests. Within days, the exchange had filed for bankruptcy, and an estimated $8 billion in customer funds was missing.

Applying the Regulatory Framework

The FTX collapse illustrates failures across virtually every dimension of the regulatory framework discussed in Chapters 3 and 4.

On the securities and commodities side, FTX offered products — leveraged tokens, derivatives — that almost certainly required registration with U.S. regulators. The company's offshore structure was explicitly designed to circumvent those requirements. The fact that FTX US, the nominally compliant U.S. entity, was capitalized with customer funds from the offshore exchange demonstrated that even the regulatory perimeter around U.S. operations was largely fictitious.

On the AML and KYC side, the evidence emerging from the bankruptcy and criminal proceedings suggested that FTX's compliance program was inadequate for an exchange of its size and customer base. The offshore structure reduced FinCEN's direct reach, but FTX US had Bank Secrecy Act obligations that appear to have been treated as formalities rather than substantive requirements.

Most fundamentally, the FTX collapse was a governance failure. The absence of independent oversight, the commingling of customer and corporate funds, the undisclosed conflicts of interest between FTX and Alameda, and the concentration of decision-making authority in a small group of insiders without meaningful checks created the conditions for fraud. No amount of trading regulations would have prevented the collapse if the people running the company were willing to steal customer funds — but the absence of governance structures that would have detected or deterred that theft made the fraud both possible and, for too long, invisible.

Bankman-Fried was convicted in 2023 on multiple counts of fraud and conspiracy and sentenced to 25 years in federal prison. Several of his associates, including the co-founders of Alameda Research and FTX's chief technology officer, pleaded guilty and cooperated with prosecutors. The criminal proceedings provided a detailed public record of the misconduct — and a clear illustration of what the regulatory frameworks that FTX evaded were designed to prevent.

Lessons for Regulatory Analysis

When analyzing the FTX collapse, several analytical frameworks from this and the preceding chapters are directly applicable.

The regulatory arbitrage framework from Chapter 1 explains why FTX chose an offshore structure: to access global markets while minimizing compliance obligations. The same pattern appears in the Binance case discussed below, and in the broader phenomenon of crypto companies structuring their operations to avoid the jurisdiction of the regulators most likely to scrutinize them.

The governance failure framework — the relationship between inadequate internal controls, absent oversight, and fraud — will be developed in detail in Chapters 11 and 12. FTX is a case study in what happens when those controls are not merely weak but deliberately absent. The fact that Bankman-Fried publicly cultivated an image of responsible crypto leadership while privately operating a fraudulent enterprise illustrates the difference between the appearance of compliance and its substance.

The consumer harm framework — who was affected, how badly, and what protections were absent — is central to evaluating what regulatory reforms might have prevented the collapse. Millions of customers lost funds they believed were safely held in custody. The absence of the customer asset segregation requirements that apply to licensed U.S. brokers and exchanges meant there was no structural protection preventing FTX from using customer funds for its own purposes.

Ripple and the XRP Litigation: Classifying Crypto Assets in Court

While the FTX collapse dominated headlines, the legal battle between the SEC and Ripple Labs over the classification of XRP represents perhaps the most consequential regulatory dispute in the history of U.S. cryptocurrency law. Unlike FTX — which involved fraud — the Ripple case involved a genuine legal question: whether a cryptocurrency that had been sold to investors over several years was a security, and what the consequences of that classification should be.

Background: What XRP Is and How It Was Sold

Ripple Labs created XRP as a digital payment token designed to facilitate fast and inexpensive cross-border money transfers. Unlike Bitcoin, which is mined by network participants, XRP was created in its entirety at inception, with Ripple Labs retaining a large portion and selling the remainder to investors and on the open market over time. Ripple Labs used XRP sales as a primary source of funding for its operations — selling billions of dollars worth of XRP over several years.

The SEC filed suit against Ripple Labs in December 2020, alleging that XRP was a security and that its sale constituted an unregistered securities offering. The complaint alleged that Ripple Labs had raised over $1.3 billion through XRP sales without registering the offering with the SEC or qualifying for an exemption — and that the company's principals had personally profited from those sales.

The Court's Analysis and Its Significance

The Ripple litigation produced a significant judicial ruling in 2023 that drew a distinction that the industry had long argued for but that no court had clearly established. Judge Analisa Torres of the Southern District of New York held that XRP sales to institutional investors — sales made through private contracts with sophisticated buyers who had access to Ripple's disclosures about the company and its use of proceeds — did satisfy the Howey Test and constituted unregistered securities offerings.

However, the court also held that XRP sales on public cryptocurrency exchanges — so-called programmatic sales where buyers purchased XRP without any direct relationship with Ripple Labs and often without knowing that the seller was Ripple — did not satisfy the Howey Test, because those buyers had no reasonable expectation that Ripple's efforts would generate their profits. They were buying a digital asset on an open market, not investing in a common enterprise with Ripple.

This distinction — between institutional sales that are securities and open-market sales that are not — had enormous implications for the industry. It suggested that the same asset could be a security in some contexts and not in others, depending on the circumstances of the sale and the reasonable expectations of the buyer. The SEC appealed aspects of the ruling, and the broader legal question of how the Howey Test applies to secondary market crypto trading remained contested as of early 2026.

The Ripple case also illustrated the practical consequences of securities classification for a company that had been selling an unregistered security for years. Even a partial victory for the SEC — the finding that institutional sales were unregistered securities offerings — exposed

Ripple to significant penalties and the possibility of disgorgement of the proceeds from those sales. The company ultimately reached a settlement with the SEC in 2024, agreeing to pay a civil penalty and accepting certain restrictions on its XRP sales, while the broader question of XRP's status in secondary markets remained unresolved.

What Ripple Teaches About Classification

The Ripple litigation reinforces several lessons that Chapter 3 introduced. First, the classification of a crypto asset as a security is not determined by the name of the asset or the intentions of its creator — it is determined by the economic reality of how it is sold and what buyers reasonably expect. The same XRP token was a security in one context and not in another, because the circumstances of the sale affected whether the Howey elements were satisfied.

Second, the consequences of getting the classification wrong are severe and long-lasting. Ripple had been selling XRP for years before the SEC's lawsuit — years during which buyers had accumulated XRP on the reasonable assumption that it was a commodity or currency, not a security. The retroactive application of securities law to those sales created uncertainty about the legal status of every XRP holder's position and complicated the path to settlement.

Third, the Ripple case highlights the absence of a clear legislative framework for crypto classification as a structural problem. Had Congress established clear criteria for when a digital asset is a security versus a commodity, Ripple might have had a defined pathway to compliance — or a clear answer that its token was not a security — rather than spending years in litigation. The CLARITY Act proposals discussed later in this chapter are a direct legislative response to exactly this problem.

Binance and the Limits of Offshore Structures

Binance is the world's largest cryptocurrency exchange by trading volume. Founded in 2017 by Changpeng Zhao, it grew rapidly by offering a wide range of crypto products, low fees, and a permissive onboarding process that did not initially require extensive identity verification. Its approach to regulation was, for most of its history, to avoid it — moving its nominal headquarters from country to country, declining to establish a principal place of business in any jurisdiction with robust financial regulation, and explicitly marketing its services to U.S. customers while maintaining that it was not subject to U.S. law.

The Pattern of Regulatory Evasion

Binance's regulatory strategy was, in retrospect, a textbook example of regulatory arbitrage taken to its extreme. The company's leadership explicitly discussed, in internal communications later revealed in court proceedings, strategies for reducing its regulatory exposure in the United States while continuing to serve U.S. customers. This included advising U.S. customers on how to circumvent geographic restrictions, routing transactions through structures designed to obscure their connection to U.S. persons, and delaying the implementation of KYC requirements that would have identified U.S. customers.

The AML failures at Binance were extensive and well-documented in the government's case against the company. Binance processed transactions for customers in Iran, Cuba, Syria, and other sanctioned jurisdictions. It facilitated transactions involving known terrorist financing typologies. Its transaction monitoring systems were, according to the government's allegations, deliberately calibrated to minimize the number of suspicious activity reports it was required to file — the opposite of a compliance program designed in good faith.

The Settlement and Its Implications

In November 2023, Binance and its founder Changpeng Zhao pleaded guilty to federal criminal charges, including violations of the Bank Secrecy Act and failure to maintain an effective AML program. Binance agreed to pay approximately $4.3 billion in penalties — one of the largest corporate penalties in U.S. history — and to submit to monitoring by an independent compliance monitor for several years. Zhao personally pleaded guilty to a charge of failing to maintain an effective AML program and was sentenced to four months in federal prison.

The Binance case established several important principles for the industry. First, operating outside U.S. borders does not protect a crypto exchange from U.S. law enforcement if the exchange serves U.S. customers. The reach of the Bank Secrecy Act, the sanctions laws administered by the Office of Foreign Assets Control, and the anti-money laundering statutes extends to foreign entities that conduct business with U.S. persons. Second, the failure to implement genuine KYC and AML programs — as opposed to superficial compliance theater — exposes a company to criminal liability, not merely civil penalties. Third, the personal liability of senior executives for compliance failures is real. Zhao's guilty plea and prison sentence demonstrated that the government was willing to pursue individual accountability, not just corporate fines.

The Binance settlement also illustrated the role of AML and KYC requirements as tools of broader national security and law enforcement policy. The government's case was not merely about regulatory compliance — it was about Binance's role in facilitating transactions for terrorist organizations and sanctioned states. Crypto exchanges that dismiss AML and KYC as bureaucratic burdens miss the fundamental point: these requirements exist because financial institutions, including crypto exchanges, are potential conduits for activity that causes serious harm.

Celsius Network and the Collapse of Crypto Lending

The collapses of Celsius Network and several similar crypto lending platforms in 2022 illustrated a different dimension of regulatory failure — one rooted not in outright fraud, as in the FTX case, but in the application of banking-like business models to an unregulated sector.

The Business Model and Its Risks

Celsius Network offered customers the ability to deposit cryptocurrency and earn interest on their holdings — in some cases, interest rates exceeding ten percent annually at a time when traditional savings accounts were earning fractions of a percent. The pitch was straightforward: deposit your crypto with Celsius, and Celsius will lend it out to institutional borrowers, earning a spread that allows it to pay depositors attractive yields.

This is, in structure, essentially what banks do. Banks take deposits, lend them out at higher interest rates, and pay depositors a portion of the spread. But banks operate under a comprehensive regulatory framework designed specifically to manage the risks of this business model: capital requirements ensure that a bank can absorb losses without failing; liquidity requirements ensure that it can meet withdrawal demands even under stress; regular examinations allow regulators to identify problems before they become crises; and deposit insurance protects customers if the bank fails anyway.

Celsius had none of these protections. It was not a bank. It was not subject to capital or liquidity requirements. It was not regularly examined by any regulator with authority over its core business. When the crypto markets fell sharply in mid-2022, the value of the collateral backing Celsius's loans collapsed, and the company found itself unable to meet customer withdrawal requests. It froze customer accounts in June 2022 and filed for bankruptcy the following month. Customers who had deposited cryptocurrency expecting bank-like safety discovered that their funds were unsecured claims in a bankruptcy proceeding.

The Regulatory Gap and Its Consequences

Celsius and similar platforms — Voyager Digital, BlockFi, and others that failed around the same time — operated in a regulatory gap that was both well-known and largely unaddressed before the collapses. State securities regulators in several states had issued cease-and-desist orders against Celsius's interest-bearing accounts, arguing that the accounts were unregistered securities — a yield-bearing investment product sold to retail customers. Some states reached settlements with Celsius requiring it to register its products. But federal regulation of the core business model — taking deposits and lending them out — was absent.

The Celsius collapse prompted serious discussion among policymakers about whether crypto lending platforms should be required to operate under bank-like regulatory frameworks. If a platform is taking customer deposits and lending them out to generate yield, the argument goes, it is performing the economic function of a bank and should face the regulatory obligations that banks face, including capital requirements, liquidity standards, and regular examination.

This analysis connects directly to the broader theme of Chapters 1 and 2: the traditional regulatory framework is built around institutions, and companies that perform institutional functions without institutional licenses can create consumer harm that existing frameworks are not designed to prevent. Celsius is the crypto lending version of the same problem that unregulated mortgage originators presented in the years before the 2008 financial crisis — a business model performing banking functions, serving retail customers, and operating outside the regulatory perimeter designed to manage exactly these risks.

The Current Direction of U.S. Crypto Regulation

The regulatory environment for cryptocurrency in the United States has shifted meaningfully since 2024, reflecting both changes in agency leadership and a broader recalibration of the political and policy consensus around crypto regulation. Understanding this shift — and its limits — is essential for any professional working in or around the crypto sector.

The Post-2024 SEC Posture

For most of the period from 2017 through 2024, the SEC's approach to cryptocurrency was primarily enforcement-driven. The agency brought dozens of actions against crypto companies for selling unregistered securities, operating unregistered exchanges, and engaging in fraud and manipulation. It declined to approve spot Bitcoin exchange-traded funds for years, citing concerns about market manipulation. It pursued high-profile cases against major exchanges and took the position that most crypto assets other than Bitcoin were securities subject to SEC jurisdiction.

Under new leadership installed after the 2024 elections, the SEC began withdrawing or settling several of its pending enforcement actions against crypto companies, including cases that had been closely watched by the industry as potential precedent-setters. The agency created a dedicated crypto task force with a mandate to develop clearer regulatory guidance for the industry rather than continuing to rely primarily on enforcement as the mechanism for establishing standards. The SEC also approved spot Bitcoin ETFs in January 2024 — a decision that had been refused multiple times under prior leadership — allowing retail investors to gain exposure to Bitcoin through familiar brokerage accounts without directly holding the asset.

These changes were significant, but they did not alter the underlying legal framework. The Howey Test remains the standard for determining whether a crypto asset is a security. The registration requirements of the Securities Act and the Securities Exchange Act still apply to offerings and platforms that fall within their scope. What changed was the vigor and direction of enforcement, not the law itself. Companies that assumed the post-2024 environment meant they could sell tokens freely without regard to securities law were misreading the situation.

The post-2024 posture also reflected a genuine policy debate about whether the SEC's prior approach had been counterproductive — driving crypto businesses offshore rather than bringing them into the U.S. regulatory perimeter. If a company faces an existential enforcement risk for operating in the United States, it will operate elsewhere. The result may be that U.S. consumers use foreign platforms with less oversight and fewer consumer protections than they would have if regulated U.S. platforms were available. This argument is not a justification for abandoning investor protection, but it is a legitimate consideration in calibrating the aggressiveness of enforcement.

Stablecoin Legislation

One area where legislative progress has been more concrete than on the broader crypto classification question is stablecoins. Stablecoins — cryptocurrencies designed to maintain a stable value, typically pegged to the U.S. dollar — have attracted regulatory attention for several reasons. They are widely used within crypto markets as a means of holding value without converting back to fiat currency. They have potential applications as a payment mechanism that could challenge the dominance of traditional payment networks. And the failure of the TerraUSD algorithmic stablecoin in May 2022 — which lost its dollar peg and collapsed to near zero within days, wiping out tens of billions of dollars in value — demonstrated that stablecoins could pose systemic risks if widely adopted.

Congressional proposals for stablecoin legislation have focused on several key issues: requiring stablecoin issuers to maintain full reserve backing in high-quality liquid assets; establishing federal licensing requirements for stablecoin issuers; requiring regular audits and public disclosure of reserve holdings; and determining whether stablecoin oversight should sit primarily with federal banking regulators, the SEC, or a new regulatory body. As of early 2026, comprehensive stablecoin legislation had advanced further in the legislative process than broader crypto market structure legislation, reflecting both the urgency of the stablecoin policy question and the relative ease of reaching consensus on the core principles.

The CLARITY Act and the Future of Crypto Regulation

The most comprehensive legislative proposal for addressing the fundamental regulatory uncertainty in the cryptocurrency sector is the framework that has been developed under various names and in various forms in Congress, most commonly referred to as the CLARITY Act or the Digital Asset Market Structure bill. Understanding what this legislation proposes, why it matters, and what its limitations are is essential for evaluating the direction of crypto regulation.

What the CLARITY Act Proposes

At its core, the CLARITY Act framework attempts to do what decades of enforcement action and agency guidance have failed to accomplish: establish clear statutory criteria for determining when a digital asset is a security subject to SEC oversight and when it is a digital commodity subject to CFTC oversight. The legislation would do this by creating a new category — the digital commodity — for crypto assets that function primarily as a medium of exchange or store of value on a sufficiently decentralized network, and placing those assets under CFTC jurisdiction with a regulatory framework tailored to the specific characteristics of digital commodity markets.

Assets that do not meet the criteria for digital commodity classification — those that are sold with the expectation of profit from the efforts of an identifiable development team, for example — would remain securities subject to SEC oversight. The legislation would also create a pathway for assets that are initially classified as securities to transition to commodity status as their networks become more decentralized, addressing the question that the Hinman speech raised informally and that the courts have not clearly resolved.

The table below summarizes how the CLARITY Act framework would address the key regulatory gaps that the current enforcement-based approach has failed to resolve.

Issue	Current State	Under CLARITY Act Proposals
Which agency has jurisdiction over a crypto asset?	Determined case-by-case through enforcement; SEC and CFTC have overlapping and contested claims	Statutory framework would establish whether an asset is a digital commodity (CFTC) or a digital security (SEC) based on defined criteria
Can a security become a commodity over time?	No clear legal pathway; addressed only informally through the non-binding Hinman speech	Decentralization pathway would allow networks to transition from SEC to CFTC jurisdiction as they mature
What rules apply to crypto exchanges?	Unclear; SEC argues most are unregistered securities exchanges; many operate in regulatory gray area	Exchanges would register with the appropriate agency based on the assets they list; separate registration categories for digital commodity and digital security platforms
How are stablecoins treated?	No comprehensive federal framework; patchwork of agency guidance and state law	Dedicated stablecoin provisions would establish reserve requirements, issuer standards, and federal oversight
What consumer protections apply?	Varies by asset classification and platform type; significant gaps for retail crypto investors	Disclosure requirements, customer asset segregation rules, and complaint mechanisms would apply across platforms

Arguments in Favor of the CLARITY Act Framework

Proponents of the CLARITY Act framework argue that the current situation — in which crypto regulation is determined primarily through enforcement actions, with the SEC and CFTC both claiming jurisdiction over assets that neither agency was designed to regulate — is untenable for a sector that now involves tens of millions of U.S. investors. A clear statutory framework would allow companies to make business decisions based on known rules rather than trying to predict the outcome of enforcement actions. It would allow legitimate crypto businesses to operate in the United States rather than seeking offshore jurisdictions to avoid regulatory uncertainty. And it would create a level playing field in which all participants operate under the same rules, rather than a system in which some companies comply and others evade.

The consumer protection argument for legislative clarity is also significant. Regulatory uncertainty does not protect consumers — if anything, it exposes them to greater risk by leaving a large and growing market effectively ungoverned. The FTX collapse and the Celsius failure both occurred in the absence of the kind of structural requirements — customer asset segregation, capital requirements, regular examination — that a clear statutory framework would likely impose. Bringing crypto platforms within a defined regulatory perimeter, with enforceable standards and meaningful oversight, may provide more consumer protection than the current enforcement-after-the-fact approach.

Arguments Against and Areas of Concern

Critics of the CLARITY Act framework raise several concerns. The most fundamental is that the legislation may effectively remove from SEC oversight assets that should remain subject to investor protection rules. If a token was initially sold as a security — with investors providing capital in expectation of profit from the development team's efforts — the fact that the network later becomes more decentralized does not retroactively make those initial investors whole or undo the harm from any fraud in the original offering. A decentralization pathway that allows securities issuers to escape SEC oversight may create perverse incentives for companies to conduct fraudulent initial offerings and then claim decentralization as a defense.

Consumer advocates have also raised concerns about whether CFTC regulation of digital commodity markets would provide adequate protection for retail investors. The CFTC's historical mandate has been focused on professional and institutional participants in derivatives markets, not on retail consumers buying assets for investment. Applying a framework designed for sophisticated market participants to the retail crypto investor buying Bitcoin on a mobile app may leave significant gaps in consumer protection.

The legislative process itself has been complicated by disagreements between the SEC and CFTC over jurisdictional boundaries, by lobbying from both the crypto industry and the traditional financial sector, and by the broader political environment around financial regulation. As of early 2026, comprehensive digital asset market structure legislation had not been enacted, though the prospects for passage had improved compared to prior years.

What Regulatory Clarity Would and Would Not Solve

It is important to be clear about what legislative clarity would and would not accomplish. A well-designed CLARITY Act framework would resolve the jurisdictional question — which agency oversees which assets — and establish the registration and disclosure requirements applicable to crypto platforms. It would not prevent all fraud. Sam Bankman-Fried stole customer funds not because no regulatory framework applied to FTX but because he chose to commit fraud. Regulation creates incentives and constraints, but it does not prevent all bad behavior by determined wrongdoers.

What regulation does provide is a structure within which fraud is more likely to be detected before it reaches catastrophic scale. Customer asset segregation requirements mean that the mixing of customer and company funds — the core of the FTX fraud — would be immediately visible to regulators conducting routine examinations. Capital requirements mean that an exchange that is losing money cannot simply draw on customer funds to stay afloat. Regular examination means that governance failures are more likely to be identified while there is still time to correct them. These structural protections do not make fraud impossible, but they make it harder to sustain and more likely to be caught.

Patterns of Failure: A Framework for Analysis

Looking across the cases examined in this chapter — FTX, Ripple, Binance, Celsius — several common patterns emerge that are useful frameworks for analyzing any crypto company that faced regulatory challenges or ethical failures.

Regulatory Arbitrage

The most consistent pattern across major crypto failures is the deliberate use of regulatory arbitrage — structuring corporate operations to minimize regulatory oversight while maximizing market access. FTX was incorporated in the Bahamas to avoid U.S. exchange registration requirements. Binance declined to establish a principal place of business in any regulated jurisdiction for years. Celsius operated as a financial intermediary without the banking license that would have subjected it to the regulatory framework designed for exactly its business model.

Regulatory arbitrage is not always illegal — companies are generally free to structure their operations in ways that minimize compliance obligations, within limits. But in the crypto context, regulatory arbitrage has repeatedly been used not merely to reduce compliance costs but to avoid the substantive requirements — customer asset protection, capital adequacy, AML compliance — that exist to prevent consumer harm. When regulatory arbitrage is used to evade requirements that exist for good reasons, the predictable result is the kind of consumer harm that occurred at FTX and Celsius.

Governance Failures

Every major crypto collapse has involved serious governance failures — the absence of independent oversight, conflicts of interest between insiders, inadequate internal controls, and concentration of decision-making authority without meaningful checks. FTX had no independent board. Celsius's management made risky investment decisions without adequate risk management infrastructure. Binance's compliance function was, according to the government's case, subordinate to business development considerations rather than operating as an independent check on conduct.

These governance failures are not unique to crypto — they appear in financial fraud cases across industries. What makes them particularly consequential in crypto is the combination of rapid growth, limited regulatory oversight, and assets that are easily moved and difficult to recover once misappropriated. Governance structures that might be adequate for a small startup become dangerously inadequate at the scale of a major exchange handling billions of dollars in customer assets.

The Gap Between Presentation and Reality

A third pattern is the gap between how crypto companies presented themselves — as innovative, responsible actors building the future of finance — and the reality of their operations. Bankman-Fried was a prominent figure in discussions about responsible crypto regulation at the same time he was operating a fraudulent enterprise. Celsius marketed itself as offering bank-like safety for customer deposits while operating with none of the protections that actually make banks safe. Binance cultivated a compliance function that appeared robust from the outside while, according to the government's case, deliberately undermining its effectiveness from within.

This pattern is relevant for professional practice. Evaluating a crypto company's regulatory and governance posture requires looking beyond public statements and marketing materials to the substance of the company's actual practices. Red flags that should prompt deeper scrutiny include: offshore incorporation in jurisdictions with light financial regulation; vague or inconsistent explanations of how customer funds are held; yield promises that significantly exceed market rates without a clear and credible explanation of the source of those returns; the absence of audited financial statements; and governance structures that concentrate authority in a small group of founders without independent oversight.

Key Takeaways

AML and KYC requirements apply to cryptocurrency exchanges as money services businesses under the Bank Secrecy Act. These requirements include customer identity verification, ongoing due diligence, suspicious activity reporting, and travel rule compliance for fund transfers above threshold amounts. The failure to implement genuine AML and KYC programs — as opposed to superficial compliance theater — exposes exchanges to criminal liability, not merely civil penalties, as the Binance case demonstrated.

The FTX collapse was caused by the misappropriation of customer funds to support the trading activities of an affiliated firm, Alameda Research, in the absence of any meaningful governance, internal controls, or regulatory oversight. The case illustrates the consequences of regulatory arbitrage, governance failure, and the absence of customer asset protection requirements. Bankman-Fried's conviction and 25-year sentence established personal criminal accountability for the founders and executives responsible.

The Ripple litigation established that the same crypto asset can be treated as a security in some contexts — direct sales to institutional investors with an expectation of profit from Ripple's efforts — and not in others — open-market trading by buyers with no direct relationship with the issuer. This context-dependent analysis reflects the application of the

Howey Test to novel circumstances and highlights the importance of the specific circumstances of each transaction in determining its regulatory treatment.

The Binance settlement demonstrated that offshore structures do not insulate a crypto exchange from U.S. law enforcement if the exchange serves U.S. customers, and that personal liability for senior executives is a real consequence of compliance failures. The $4.3 billion penalty and Changpeng Zhao's prison sentence are the most significant enforcement outcomes in the history of crypto AML regulation.

The Celsius collapse illustrates what happens when a platform performs the economic function of a bank — taking deposits and lending them out to generate yield — without the regulatory framework that makes banks safe. The absence of capital requirements, liquidity standards, and regular examination allowed Celsius to take risks with customer funds that no regulated bank would have been permitted to take.

The post-2024 SEC enforcement posture represents a meaningful shift in approach — from aggressive enforcement of securities laws across the crypto sector to a more selective and dialogue-oriented stance — but does not change the underlying legal framework. The Howey Test remains the standard for determining whether a crypto asset is a security.

The CLARITY Act framework proposes to resolve the fundamental regulatory uncertainty in the crypto sector by establishing statutory criteria for distinguishing digital commodities from digital securities, creating a decentralization pathway for networks to transition from SEC to CFTC jurisdiction, and establishing registration and disclosure requirements for crypto exchanges. As of early 2026, comprehensive legislation had not been enacted, and the regulatory environment remains defined primarily by enforcement history and agency guidance.

The common patterns across major crypto failures — regulatory arbitrage, governance failure, and the gap between public presentation and operational reality — provide a practical framework for analyzing any company that faced regulatory challenges or ethical failures. These patterns are not unique to crypto, but they are especially consequential in a sector characterized by rapid growth, significant retail participation, and historically limited oversight.

Chapter 5: Banking, Payments, and the FinTech Challenge to Legacy Systems

Introduction

Every time you tap your phone to pay for coffee, receive a direct deposit from your employer, or transfer money to a friend, you are using one of the most complex and consequential pieces of infrastructure in the modern economy: the payment system. Payment systems are the plumbing of finance. They are invisible when they work — and catastrophic when they do not. They move trillions of dollars every day, link every bank account in the country to every other, and underpin virtually every economic transaction that takes place.

For most of the twentieth century, this plumbing was built and owned by banks. The Automated Clearing House network, wire transfer systems, and card payment networks were designed by and for traditional financial institutions. The rules governing them were written with those institutions in mind. When FinTech companies arrived with faster, cheaper, and more user-friendly payment products, they encountered a system that was not designed for them — and a regulatory framework that had never contemplated their existence.

That collision between FinTech innovation and legacy payment infrastructure is the subject of this chapter. It covers how traditional payment systems work, how they are regulated, and why they are both enormously durable and increasingly under pressure from new entrants. It covers the FinTech alternatives — mobile wallets, peer-to-peer platforms, real-time payment networks — and what they mean for consumers, banks, and regulators. And it covers the question that sits at the intersection of payments and banking regulation: what options do

FinTech companies have when they want to become banks, or behave like banks, and what are the regulatory consequences of each path?

This chapter contains a significant amount of factual content that students are expected to know with precision. The quiz topics for Module 3 are specific — particular payment systems, particular charter types, particular legal standards — and this chapter is where those topics are taught. Read carefully and pay attention to the distinctions that matter: ACH versus wire transfer, ILC versus special purpose charter, Visa as a network versus Visa as an issuer. These distinctions show up in assessments and, more importantly, in professional practice.

The Purpose of a Payment System

Before examining specific payment systems, it is worth establishing what any payment system is designed to do. The ultimate purpose of any payment system is settlement — the final, irrevocable transfer of value from one party to another. Everything else in the payment process — authorization, clearing, message routing, fraud screening — exists to facilitate settlement. When settlement occurs, the transaction is complete. The payer's account has been reduced, the payee's account has been increased, and the obligation between them is discharged.

Settlement is not instantaneous in most payment systems, even when the transaction feels instant to the consumer. When you swipe a debit card at a grocery store, the authorization — the check that you have sufficient funds and the approval of the transaction — happens in seconds. But the actual movement of funds between the bank that issued your card and the bank that serves the grocery store may take one to three business days. During that interval, the funds are in transit — committed but not yet fully settled. This gap between authorization and settlement is a source of risk, cost, and regulatory attention across payment systems.

Understanding settlement also helps explain why payment system regulation exists. The risk that a party to a settlement fails before the transaction clears — called settlement risk — can cascade through the financial system if the transaction volumes are large enough. This is why the Federal Reserve, which operates the major wholesale payment systems in the United States, imposes risk controls on participants, requires intraday credit to be collateralized, and monitors settlement flows in real time. It is also why the entry of new, lightly regulated participants into the payment system raises supervisory concern: a platform that processes billions of dollars in transactions daily without the capital buffers and risk controls that apply to banks can create systemic exposure that extends well beyond the platform itself.

Traditional Payment Systems

The Automated Clearing House Network

The Automated Clearing House network, universally known as ACH, is the most prevalent retail interbank funds transfer system in the United States. It processes direct deposits, payroll payments, Social Security benefit disbursements, mortgage payments, utility bill payments, subscription charges, and an enormous range of other consumer and business transactions. In a typical year, the ACH network processes tens of billions of transactions representing tens of trillions of dollars in value.

ACH was created in the early 1970s as an electronic alternative to paper checks. Its architecture reflects that origin: it is a batch processing system, meaning transactions are not processed one at a time in real time but rather are collected, sorted, and transmitted in batches at scheduled intervals throughout the business day. The network is operated jointly by the Federal Reserve's FedACH service and by The Clearing House's EPN (Electronic Payments Network), a private-sector operator. The rules governing ACH are established by NACHA — the National Automated Clearinghouse Association — a self-regulatory organization whose rules are binding on all financial institutions that participate in the network.

An ACH transaction begins when an originator — an employer, a biller, or any entity initiating a payment — sends a payment instruction to its bank, known as the Originating Depository Financial Institution, or ODFI. The ODFI transmits the transaction to the ACH operator, which sorts and routes it to the Receiving Depository Financial Institution, or RDFI — the bank that holds the recipient's account. The RDFI posts the credit or debit to the recipient's account and the transaction settles, typically within one business day for standard ACH or on the same business day for Same-Day ACH, which NACHA implemented in phases beginning in 2016.

ACH is inexpensive — processing costs are fractions of a cent per transaction for participating financial institutions — which makes it the workhorse of consumer and business payments in the United States. Its primary limitation is speed: despite the availability of Same-Day ACH, the system still operates on business day cycles and does not process on weekends or federal holidays. This limitation has become increasingly significant as consumers accustomed to instant everything have come to expect immediate availability of funds. It has also created a market opportunity for faster alternatives, discussed later in this chapter.

From a regulatory perspective, ACH participants are subject to NACHA's operating rules, which establish the responsibilities of ODFIs and RDFIs, set return timeframes for failed transactions, and govern the handling of unauthorized debits. Federal oversight comes from the Federal Reserve in its capacity as both an ACH operator and a supervisor of the banking institutions that participate in the network. The CFPB has enforcement authority over consumer protection aspects of ACH transactions, particularly the Electronic Fund Transfer Act, which governs consumer rights in electronic payment disputes.

Wire Transfers

Wire transfers are the payment system used for large-value, time-critical transfers that require same-day finality. While ACH processes the volume of U.S. payments, wire transfer systems process the value — the daily dollar amounts moving through Fedwire, the Federal Reserve's wire transfer system, regularly exceed ten trillion dollars.

The primary wire transfer systems in the United States are Fedwire Funds Service, operated by the Federal Reserve, and CHIPS — the Clearing House Interbank Payments System — operated by The Clearing House and used primarily for large interbank and international dollar transactions. Fedwire provides real-time gross settlement, meaning each transaction is processed and settled individually and immediately rather than being netted against other transactions. CHIPS uses a multilateral netting mechanism that reduces the total amount of funds that must actually move to settle a day's transactions, improving liquidity efficiency for participants.

A wire transfer initiated by a consumer or business goes from the sender's bank — which debits the sender's account and sends a payment message through Fedwire or CHIPS — to the recipient's bank, which credits the recipient's account. Because Fedwire settles in real time, once a wire transfer is sent, it is final and irrevocable. This finality is both the great advantage of wire transfers — the recipient knows the funds will not be reversed — and a significant risk factor, because fraudulent or erroneous wire transfers cannot be recalled once processed.

Wire transfer fraud — particularly business email compromise schemes in which fraudsters impersonate executives or vendors to redirect legitimate wire transfers to fraudulent accounts — is among the most costly forms of financial fraud in the United States. The FBI's Internet Crime Complaint Center consistently reports business email compromise as one of the highest-dollar fraud categories it tracks. The combination of wire transfer finality and the social engineering techniques used in these schemes makes recovery difficult and has made wire transfer security a significant focus for both financial institutions and their corporate clients.

Wire transfers are regulated under Article 4A of the Uniform Commercial Code, which governs the rights and obligations of banks and their customers in wholesale funds transfers. Federal regulatory oversight comes from the Federal Reserve and from the banking regulators that supervise the institutions participating in wire transfer systems.

Card Payment Networks

Card payments — credit cards, debit cards, and prepaid cards — are the most visible form of electronic payment for most consumers. They are also among the most complex, involving multiple parties, multiple fee arrangements, and a regulatory framework that touches federal consumer protection law, network rules, and antitrust considerations.

Understanding how card payments work requires distinguishing between four distinct roles that are often confused: the card network, the issuing bank, the acquiring bank, and the merchant. The card network — Visa, Mastercard, American Express, Discover — is the infrastructure provider. It operates the global communication and clearing system that routes authorization requests and settlement messages between the banks on either side of a transaction. Critically, Visa and Mastercard do not issue credit cards directly. They do not have direct relationships with cardholders. They do not extend credit. They provide the network — the rules, the technology, and the brand — and they earn revenue from the fees generated by transactions flowing through that network.

The issuing bank — also called the card issuer — is the financial institution that actually issues the card to the consumer, extends credit in the case of a credit card, and is responsible for the consumer relationship. When you carry a Visa card bearing the name of your bank, your bank is the issuer. It has a license agreement with Visa to issue cards on the Visa network, but it makes the credit decisions, sets the interest rates, and handles customer disputes. The Visa branding indicates which network the card runs on, not who issued it or who is extending the credit.

The acquiring bank — also called the merchant acquirer — is the financial institution that has a relationship with the merchant and processes the merchant's card transactions. When a merchant signs up to accept Visa cards, it enters into an agreement with an acquirer. The acquirer provides the point-of-sale technology, processes the merchant's transactions, and deposits the net proceeds — after deducting fees — into the merchant's account.

The merchant pays for the privilege of accepting cards through a set of fees collectively known as the merchant discount rate. The largest component of that rate is interchange — a fee paid by the acquiring bank to the issuing bank for each transaction. Interchange rates are set by the card networks and vary based on the type of card, the type of transaction, and the type of merchant. For most consumer credit card transactions, interchange rates range from roughly one and a half to three and a half percent of the transaction value. Debit card interchange is lower and, for large issuers, is capped under the Durbin Amendment to the Dodd-Frank Act at approximately 0.05 percent plus $0.21 per transaction.

The interchange fee structure has been a source of persistent controversy. Merchants argue that the fees are excessive and opaque, forced upon them by networks with dominant market positions. Banks and card networks argue that interchange compensates issuers for the credit risk and fraud losses they bear and funds the rewards programs that consumers value. The debate has produced litigation, legislative proposals, and regulatory interventions in multiple countries. In the United States, the Durbin Amendment's debit interchange cap was the most significant federal intervention, but the broader question of credit card interchange remains active.

Consumer Protections in Card Payments

Federal law provides important consumer protections in the context of card payments. The Truth in Lending Act, implemented through Regulation Z, requires issuers to disclose the terms of credit card accounts — interest rates, fees, grace periods — in standardized formats that allow consumers to compare products and understand their costs. It also limits consumer liability for unauthorized charges and governs the dispute resolution process for billing errors.

The Fair Credit Billing Act, enacted as an amendment to TILA, gives consumers the right to dispute billing errors on credit card statements and requires issuers to investigate and respond to disputes within specific timeframes. The consumer's liability for unauthorized credit card charges is limited to $50 under TILA — and most major issuers have voluntarily adopted zero-liability policies that go beyond what the law requires.

The Credit Card Accountability Responsibility and Disclosure Act of 2009 — commonly called the CARD Act — significantly strengthened consumer protections for credit card users. It restricted interest rate increases on existing balances, prohibited over-limit fees without prior consumer opt-in, required that payments in excess of the minimum be applied to the highest-interest balances first, and established protections specifically for younger consumers. The CARD Act was a direct legislative response to the aggressive marketing and pricing practices that had become common in the credit card industry before the 2008 financial crisis.

For debit cards, the Electronic Fund Transfer Act and its implementing regulation, Regulation E, govern consumer rights. The unauthorized transaction liability framework under Regulation E differs from TILA: a consumer who reports an unauthorized debit card transaction within two business days is liable for no more than $50; reporting within 60 days limits liability to $500; failure to report within 60 days can result in unlimited liability. This framework creates meaningful incentives for consumers to monitor their accounts and report problems promptly.

The Settlement Process in Card Transactions

The typical card transaction follows a two-phase process: authorization and settlement. At the point of sale, when a consumer taps, swipes, or inserts a card, the merchant's point-of-sale system sends an authorization request through the acquirer to the card network, which routes it to the issuing bank. The issuer checks the account balance or credit limit, verifies the card details, performs fraud screening, and returns an authorization response — approved or declined — in a matter of seconds. The authorization reserves the funds but does not move them.

Settlement occurs separately, typically at the end of the business day when the merchant submits a batch of authorized transactions to the acquirer. The acquirer forwards the batch through the card network to the issuers, which debit the cardholders' accounts. Funds flow from the issuers through the network to the acquirer, which deposits them — net of fees — into the merchant's account. The full settlement cycle typically takes one to three business days from the transaction date, depending on the network, the acquirer, and the merchant's bank.

This gap between authorization and settlement is the source of several practical complications: a merchant that authorizes a transaction cannot be certain the funds will arrive until settlement clears; a consumer who overspends between authorization and settlement may encounter declined transactions or overdrafts; and fraudulent transactions that are not identified during authorization may be reversed through the chargeback process after settlement, creating losses for merchants. The chargeback — the reversal of a settled transaction — is one of the most significant sources of financial loss for e-commerce merchants and has generated an entire industry of chargeback management services.

FinTech Payment Innovations

Legacy payment systems — ACH, wire transfer, card networks — are enormously capable and deeply embedded in the financial infrastructure of the United States. They are also slow by modern standards, expensive relative to what technology now makes possible, and built around a model of scheduled batch processing and business-day settlement that feels increasingly anachronistic to consumers and businesses that expect real-time everything. FinTech payment companies have built their businesses primarily by addressing these limitations.

Mobile Wallets

Mobile wallets — Apple Pay, Google Pay, and Samsung Pay are the most widely used — are not fundamentally new payment systems. They are a new interface for existing payment systems, primarily card networks. When you pay with Apple Pay at a point-of-sale terminal, the underlying transaction runs on a Visa, Mastercard, or other card network, processed by an acquirer, settled through the same mechanisms as any other card transaction. The consumer experience is different — a tap of the phone rather than a physical card swipe — but the plumbing is the same.

What mobile wallets do differently is security. Instead of transmitting your actual card number to the merchant, mobile wallets use tokenization — they replace your card number with a device-specific token that is valid only for that device and that cannot be used to initiate fraudulent transactions elsewhere. If a merchant's payment system is breached and the tokens are stolen, the stolen data is useless outside the context of the specific device from which it was generated. This architecture has significantly reduced the card-present fraud that was endemic to magnetic stripe card technology.

Mobile wallets also collect payment data. Apple and Google know where their users shop, how often, and for how much — data that has value both for internal product development and, in some contexts, for advertising or other commercial purposes. The privacy implications of payment data collection are addressed in Chapters 7 and 8.

From a regulatory perspective, mobile wallet providers occupy an interesting middle space. Because the underlying transactions run on card networks, the primary regulatory framework is the one that applies to those networks and to the issuing banks. But mobile wallet providers are also large technology companies with their own commercial relationships, and the question of whether they should face additional regulation — particularly around their access to payment data and their leverage over merchants — has become an active policy debate in the United States and Europe.

Peer-to-Peer Payment Platforms

Peer-to-peer, or P2P, payment platforms — Venmo, Cash App, Zelle, and PayPal among the most prominent — are designed for the use case that card networks handle poorly: transferring money between individuals. Splitting a dinner check, paying a babysitter, reimbursing a friend — these transactions are awkward with cash and impractical with credit cards, because card networks were designed for consumer-to-merchant transactions, not person-to-person transfers.

P2P platforms handle this use case by creating closed-loop systems or by overlaying real-time messaging on existing banking infrastructure. Venmo and Cash App, for example, maintain internal ledgers: when you send $50 to a friend within the Venmo platform, the transfer is recorded instantly in Venmo's system, but the actual movement of funds from one bank account to another may happen later, when the recipient chooses to transfer their Venmo balance to their bank. Zelle, by contrast, is built directly on top of bank-to-bank transfers and moves funds between bank accounts in near real time without an intermediary balance.

The consumer experience is similar across these platforms — money arrives quickly and the interface is simple — but the underlying mechanics, and therefore the regulatory treatment, differ. Venmo and Cash App hold customer funds in internal accounts, making them money transmitters subject to FinCEN registration and state money transmitter licensing. Zelle, which is operated by Early Warning Services — a consortium owned by the major U.S. banks — does not hold funds itself;

it facilitates bank-to-bank transfers using each bank's existing infrastructure.

The fraud problems that have emerged on P2P platforms illustrate the consumer protection gaps created when payment technology outpaces regulatory oversight. Zelle in particular became the subject of significant Congressional and regulatory attention when it became apparent that the platform was being widely used to facilitate scams — fraudsters impersonating bank employees or other trusted parties to trick consumers into authorizing transfers that were promptly moved beyond recovery. Because Zelle transfers are authorized by the consumer, even under fraudulent pretenses, banks argued they were not obligated to reimburse losses. Consumer advocates and the CFPB argued that platforms with hundreds of millions of dollars in fraud losses had an obligation to do more. The resulting debate produced Congressional hearings, a CFPB enforcement action against the major bank operators of Zelle, and ongoing litigation — and it surfaced fundamental questions about the allocation of fraud risk in instant payment systems that have not yet been fully resolved.

The Zelle situation illustrates a framework that students should carry forward: when a FinTech platform creates conditions that facilitate consumer harm at scale, the question of regulatory responsibility is not simply whether the consumer technically authorized the transaction. It also involves whether the platform's design, fraud controls, and dispute resolution practices meet the standard of care that consumer protection law and reasonable commercial practice require. Evaluating that question requires understanding both the technical mechanics of how the platform works and the legal frameworks — particularly the Electronic Fund Transfer Act and CFPB enforcement authority — that govern it.

Real-Time Payment Networks

The limitations of ACH — batch processing, business-day settlement cycles, no weekend availability — created a market gap that both private-sector operators and the Federal Reserve have moved to fill with real-time payment infrastructure.

The Clearing House's Real-Time Payments network, known as RTP, launched in 2017 and was the first new core payment infrastructure built in the United States in more than 40 years. RTP enables bank-to-bank transfers that settle in seconds, around the clock, every day of the year including weekends and holidays. Unlike ACH, which batches and nets transactions, RTP processes each payment individually and provides immediate final settlement. Banks connected to RTP can offer their customers the ability to send and receive funds instantly — a capability that was previously available only through wire transfer at much higher cost.

The Federal Reserve entered the real-time payments market with the launch of FedNow in July 2023. FedNow is a Federal Reserve-operated instant payment system that provides the same core capability as RTP — immediate, final, 24/7 settlement — but with the backing of the central bank and, importantly, with a mandate to achieve broad participation across the full range of U.S. financial institutions, including the community banks and credit unions that have been slower to connect to the private-sector RTP network.

The coexistence of two competing real-time payment networks — RTP and FedNow — reflects a deliberate policy choice to maintain both private-sector innovation and Federal Reserve infrastructure in the payments market. The Federal Reserve's rationale for building FedNow rather than simply encouraging participation in the existing RTP network centered on access and interoperability concerns: a private-sector network owned by the largest banks might not extend the same terms and access to smaller institutions that a Fed-operated network would. Whether the existence of two networks ultimately benefits or complicates the payments ecosystem remains to be seen as both networks work to expand their reach.

Real-time payment networks have significant implications for FinTech companies. Access to real-time settlement infrastructure — which has historically been available only to chartered banks — gives bank-connected FinTechs the ability to offer instant payment products that were previously impossible. But it also raises the stakes around fraud: in an instant payment system, fraudulent transfers are immediately final and cannot be reversed. The fraud controls that are adequate for a two-day ACH cycle, where unauthorized transactions can be returned before settlement, are not adequate for a system where every transaction settles in seconds.

Cryptocurrency as a Payment System

Some cryptocurrency proponents have argued that blockchain-based payment systems represent a fundamentally different approach — one that bypasses the legacy infrastructure of banks, card networks, and the ACH entirely. Bitcoin and other cryptocurrencies can be transferred directly between parties anywhere in the world in minutes, without any bank in the middle, without any settlement delay, and without any intermediary collecting fees.

In practice, cryptocurrency payments face significant limitations as a mainstream payment mechanism. Price volatility makes Bitcoin an impractical currency for most everyday transactions — a merchant that accepts Bitcoin for a $50 purchase has no certainty what that Bitcoin will be worth when converted to dollars. Transaction fees on the Bitcoin network fluctuate with network congestion and can be high during periods of heavy use. And the regulatory and tax treatment of cryptocurrency payments — each transaction is a taxable event — creates compliance burdens that most consumers and merchants are unwilling to manage.

Stablecoins — cryptocurrencies pegged to the U.S. dollar — address the volatility problem and are increasingly used as a payment mechanism within crypto markets and, to a growing extent, in cross-border commercial transactions. The regulatory framework for stablecoin payments is evolving, as discussed in Chapter 4, but the fundamental question of whether stablecoin issuers should face bank-like requirements — reserve requirements, regular audits, federal oversight — is directly relevant to their role as payment instruments.

For purposes of understanding the U.S. payment system landscape, the most important point about crypto payments is that they remain, as of early 2026, a small fraction of overall payment volume and are most widely used for specific use cases — cross-border remittances, transactions within crypto markets, and payments in markets where traditional banking infrastructure is weak — rather than as a general-purpose alternative to card and ACH payments.

The Regulatory Landscape for Payment FinTechs

A FinTech company that wants to process payments in the United States faces a regulatory landscape that is, in some respects, more complex than the one facing a traditional bank. Banks have a single primary federal regulator and operate under a well-established charter framework. Payment FinTechs face a patchwork of federal requirements, state licensing obligations, and network rules that vary based on the specific nature of their activities.

Federal Regulatory Requirements

At the federal level, payment FinTechs are primarily regulated through the Bank Secrecy Act framework administered by FinCEN. As discussed in Chapter 4, money services businesses — including money transmitters — must register with FinCEN, implement AML programs, conduct customer due diligence, and file suspicious activity reports. Most payment FinTechs that hold or transmit customer funds qualify as money transmitters and carry these obligations.

Consumer protection obligations flow from the Electronic Fund Transfer Act and Regulation E for transactions involving consumer deposit accounts, and from the CFPB's broader authority over unfair, deceptive, or abusive acts or practices in consumer financial products. The CFPB has been active in examining the practices of large payment platforms and has brought enforcement actions when consumer protection failures — particularly around unauthorized transaction disputes — reached sufficient scale.

State Money Transmitter Licensing

State money transmitter licensing, introduced in Chapter 2, is the most operationally demanding regulatory challenge for payment FinTechs. A company that transmits money on behalf of customers must obtain a money transmitter license — or its equivalent — in each state where it operates. The requirements vary by state: some states have relatively streamlined processes; others impose substantial net worth requirements, surety bonds, and background check obligations.

The practical consequence is that a payment startup that wants to offer its services to customers across the United States must navigate up to 50 separate licensing processes simultaneously. This creates a structural advantage for companies that have already completed the licensing process and a meaningful barrier to entry for new market participants. Several regulatory reform initiatives — including the Money Transmission Modernization Act, which has been adopted in a growing number of states — have attempted to harmonize state licensing requirements and reduce the compliance burden, but the patchwork remains substantially intact as of early 2026.

Network Rules and Interchange Regulation

Payment FinTechs that operate within the card network ecosystem — issuers, processors, payment facilitators — are also subject to the rules of the card networks themselves. Visa and Mastercard each maintain extensive operating rules that govern how their networks can be used, what disclosures merchants must make, how disputes are handled, and what security standards must be met. Compliance with PCI DSS — the Payment Card Industry Data Security Standard — is mandatory for any entity that handles card data, regardless of whether any government regulation requires it.

Network rules are not government regulations, but they have the practical force of regulations for companies that depend on network access. A payment company that violates Visa's operating rules can be suspended from the network — a consequence that would be existential for most payment businesses. The networks' ability to set and enforce their own rules, combined with their dominant market positions, has made network governance a subject of antitrust scrutiny and legislative attention in the United States and internationally.

Becoming a Bank: Pathways and Trade-offs

As payment FinTechs scale, many encounter a fundamental structural problem: they are building financial services businesses on top of infrastructure they do not own, subject to compliance obligations they did not create, and dependent on bank partners whose due diligence requirements and risk appetites constrain what they can offer. The logical response for some is to ask: why not just become a bank?

The question is easier to ask than to answer. Becoming a bank offers significant advantages — access to federally insured deposits, direct participation in payment systems, a single primary federal regulator replacing a patchwork of state licenses, and the trust and credibility that comes with a federal charter. But it also imposes significant obligations — capital requirements, liquidity standards, regular examination, Community Reinvestment Act obligations, and a compliance burden that can dwarf what a non-bank FinTech faces. The decision to pursue a bank charter is one of the most consequential strategic choices a FinTech company can make, and it should be made with a clear understanding of all the pathways available and the trade-offs each involves.

The table below summarizes the major charter options available to FinTech companies seeking bank status or banking capabilities.

Pathway	Charter Type	Primary Regulator	Accepts Deposits?	Key Requirements / Constraints
Full-service national bank	National bank charter (OCC)	OCC	Yes (FDIC insured)	Full bank regulation: capital, liquidity, CRA, regular examination; holding company subject to Federal Reserve
Full-service state bank (Fed member)	State charter	State regulator + Federal Reserve	Yes (FDIC insured)	State banking law; Federal Reserve holding company rules; CRA obligations
Industrial Loan Company (ILC)	State charter (limited states)	FDIC + state regulator	Yes (FDIC insured)	Not subject to Federal Reserve BHC oversight; commercial firms can own; FDIC approval required; limited states offer charter
OCC Special Purpose FinTech Charter	Special purpose national bank (OCC)	OCC	No (no deposit-taking)	Must perform at least one core banking function; capital and liquidity requirements; financial inclusion plan; recovery/resolution plan; legal status contested

Bank partnership (no charter)	No charter held by FinTech	OCC/FDIC/Federal Reserve (regulates bank partner)	Via bank partner only	No direct regulatory obligations; subject to bank partner's due diligence; limited access to Fed payment systems; rent-a-bank scrutiny

Full-Service Bank Charters

A full-service national bank charter, issued by the Office of the Comptroller of the Currency, or OCC, allows a company to operate as a full bank — accepting deposits, extending credit, and providing a full range of banking services — under federal oversight. A state bank charter, issued by a state banking regulator, provides the same core capabilities under state law, with the relevant federal banking agency (the Federal Reserve for state members, the FDIC for state nonmembers) providing federal oversight.

Obtaining a full-service bank charter requires demonstrating to the chartering authority that the applicant has adequate capital, a qualified management team, a sound business plan, and the ability to meet the needs of the communities it will serve. The application process is lengthy — typically one to three years — and the ongoing compliance obligations are substantial. The Community Reinvestment Act requires banks to demonstrate that they are meeting the credit needs of the communities in which they operate, including low- and moderate-income communities. Capital adequacy requirements, liquidity requirements, and regular safety-and-soundness examinations are permanent features of life as a bank.

For most FinTech companies, a full-service charter is more regulatory burden than they need or want. The primary draw is the ability to accept federally insured deposits — which provides cheap, stable funding — but many FinTech business models do not require deposit-taking. The compliance infrastructure required to maintain a full-service charter would consume resources that early-stage companies need for growth, and the examination process would expose business practices to regulatory scrutiny that startups may prefer to defer.

The Industrial Loan Company Charter

The industrial loan company, or ILC, charter is a state-chartered financial institution that can accept federally insured deposits but is not subject to Federal Reserve oversight as a bank holding company. This means that a commercial company — a technology firm, a retailer, or an industrial corporation — can own an ILC without subjecting the entire enterprise to Federal Reserve supervision as it would if it owned a conventional bank.

ILCs are chartered by a limited number of states — Utah, Nevada, California, and a few others — and are supervised by both the chartering state's banking regulator and the FDIC, which provides deposit insurance. The FDIC's approval is required before an ILC can accept insured deposits, giving the FDIC effective veto authority over new ILC formations.

Walmart applied for an ILC charter in 2005, seeking to establish its own banking subsidiary that would allow it to process its own payment transactions and potentially offer financial services to its customers and employees. The application generated intense opposition from the banking industry, which argued that allowing a major retailer to own a bank would mix commerce and banking in ways that posed systemic risks and anticompetitive threats. Walmart withdrew its application in 2007 after the FDIC repeatedly extended the public comment period and the political resistance to approval became clear. The episode remains the most prominent example of commercial firm interest in the ILC charter and the political dynamics that can complicate its use.

More recently, several technology companies and FinTechs — including Square, which received an ILC charter in Utah in 2020 and operates it as Square Financial Services — have successfully obtained ILC charters as a pathway to bank status that does not require the parent company to submit its entire corporate structure to Federal Reserve oversight. The ILC remains a viable pathway for the right applicant, but it is narrow, politically sensitive, and dependent on FDIC approval.

The OCC Special Purpose FinTech Charter

The OCC's special purpose national bank charter, first proposed in 2016, represents an attempt to create a federal banking license specifically designed for FinTech companies — one that confers the benefits of a national bank charter without requiring the charter holder to accept deposits. Chapter 2 introduced this charter in the context of the regulatory landscape; this chapter examines its specific requirements in more detail.

Under the OCC's framework, a special purpose national bank charter is available to companies that perform at least one of the three core banking functions: receiving deposits, paying checks, or lending money. A company that does none of these things is not eligible for a national bank charter, special purpose or otherwise. A company that does at least one — a lender that does not accept deposits, or a payments company that processes checks — may apply.

The core banking function requirement is not merely a threshold test — it shapes the entire regulatory framework that applies to the charter holder. A special purpose bank that lends money is subject to the lending-related consumer protection requirements that apply to national banks, including Truth in Lending Act obligations and fair lending laws. A special purpose bank focused on payments is subject to the payment-related requirements applicable to national banks. The OCC applies the relevant regulatory framework based on the banking functions the charter holder actually performs.

Beyond the core banking function requirement, applicants for a special purpose FinTech charter must demonstrate several things. They must have adequate capital and liquidity relative to their risk profile — the specific requirements are tailored to the applicant's business model rather than set at a fixed level, because a company that does not accept deposits does not need the same capital cushion as one that does. They must have a financial inclusion commitment — a plan for how the company will meet the financial services needs of underserved communities, reflecting the CRA-like obligations the OCC expects of national banks. And they must have a recovery and resolution plan — a credible description of how the company would wind down its operations in an orderly manner if it became financially distressed, reducing the risk that its failure would harm customers or the broader financial system.

The legal status of the OCC's special purpose FinTech charter has been contested since it was proposed. The Conference of State Bank Supervisors filed suit arguing that the OCC lacks statutory authority to charter a non-depository institution as a national bank. Courts have reached conflicting conclusions on this question, and the charter's legal viability remained uncertain as of early 2026. The OCC has continued to develop the framework, but no FinTech company had received a special purpose FinTech charter as of early 2026, reflecting both the legal uncertainty and the practical challenges of the application process.

Bank Partnerships: The Practical Alternative

For the large majority of FinTech companies, the practical pathway to banking capabilities is not a bank charter but a bank partnership. As discussed in Chapter 2, FinTech companies partner with chartered banks to access deposit infrastructure, payment systems, and regulatory permissions that they cannot obtain independently. The bank holds the deposits, issues the accounts, and carries the regulatory responsibility; the FinTech provides the technology, the user interface, and the customer relationship.

This model has produced some of the most successful FinTech companies in the country. Chime, one of the largest neobanks in the United States, does not hold a bank charter — it partners with Bancorp Bank and Stride Bank, which hold customer deposits and provide FDIC insurance. Green Dot, another major FinTech, operates both as a bank holding company and through partnerships. The SoFi Financial platform obtained a national bank charter in 2022, one of the few FinTechs to successfully complete the transition from partnership-dependent business to chartered institution.

The bank partnership model has significant limitations, however. A FinTech company that depends on a bank partner for its core regulatory permissions is vulnerable to changes in that partner's risk appetite, regulatory standing, or business strategy. If the partner bank faces regulatory scrutiny over its FinTech partnerships — as several banks serving as partners for multiple FinTechs have — it may tighten its due diligence requirements, restrict the products the FinTech can offer, or terminate the relationship entirely. The collapse of Synapse, a middleware platform that connected FinTechs to their banking partners, in 2024 left tens of thousands of customers temporarily unable to access their funds and illustrated the fragility of a model in which customers' money flows through multiple layers of unregulated intermediaries between the FinTech's app and the chartered bank holding the deposits.

The Synapse collapse is a case study in the risks of the bank partnership model at scale. It also illustrates why regulators have been increasingly attentive to the relationships between banks and their FinTech partners — and why bank due diligence requirements, the source of delays and frustration for many FinTechs seeking partnerships, exist for reasons that go beyond bureaucratic caution.

The Evolution vs. Revolution Question

A question worth addressing directly, because it frames a recurring debate in the industry: are FinTech payment systems an evolution of existing banking rails, or a revolution that is making banks less relevant?

The honest answer is both, depending on the layer you are examining. At the infrastructure layer, most FinTech payment innovation is evolutionary — it builds on top of the ACH network, the card networks, and the Federal Reserve's payment systems rather than replacing them. Venmo's instant transfer feels revolutionary to the user but is, beneath the surface, an ACH transaction. Apple Pay feels like a new technology but runs on the same card network rails as a physical swipe. FedNow is new infrastructure, but it is new infrastructure built and operated by the same central bank that has operated the country's payment systems for over a century.

At the customer relationship layer, however, the change is more genuinely disruptive. A consumer whose financial life centers on a Cash App account, a Chime debit card, and a Robinhood investment account may have no meaningful relationship with a traditional bank at all. The banks behind those products — the chartered institutions holding the deposits and providing the regulatory infrastructure — are invisible to the consumer. The FinTech company owns the customer relationship, the data, and the product experience. This is a significant shift in the structure of the financial services industry, even if the underlying infrastructure remains largely unchanged.

From a regulatory perspective, the evolution-vs.-revolution question matters because it determines what regulatory framework should apply. If FinTech payment platforms are essentially the same as bank services delivered through a different interface, they should face roughly the same regulatory requirements. If they are genuinely different — operating on fundamentally different business models, serving different customer populations, creating different risks — the appropriate regulatory framework may also need to be different. The answer, consistent with the principle established in Chapter 2, is that there is no one-size-fits-all answer. Regulatory treatment should be calibrated to the actual risks created by the actual activities performed, not to the label attached to the company performing them.

Key Takeaways

The ultimate purpose of any payment system is settlement — the final, irrevocable transfer of value from one party to another. Everything else in the payment process exists to facilitate settlement, and the risk that a party fails before settlement occurs is a central concern of payment system regulation.

ACH is the most prevalent retail interbank funds transfer system in the United States. It is a batch processing system that settles on a scheduled cycle — same business day for Same-Day ACH — and is governed by NACHA operating rules alongside Federal Reserve oversight. Its low cost and ubiquity make it the workhorse of consumer and business payments; its batch processing architecture is its primary limitation.

Wire transfers, processed through Fedwire and CHIPS, provide real-time gross settlement for large-value transfers. They are final and irrevocable, which makes them highly reliable and highly vulnerable to fraud. Business email compromise schemes that redirect wire transfers are among the highest-dollar fraud categories in the United States.

Visa and Mastercard are card networks — they do not issue credit cards directly. Issuing banks issue the cards, extend credit, and maintain the consumer relationship. Acquiring banks serve merchants. Card networks provide the infrastructure and rules that connect them. Interchange fees, paid by acquiring banks to issuing banks, are the largest component of the merchant discount rate and have been a persistent source of regulatory controversy.

Consumer protections in card payments are provided primarily by the Truth in Lending Act (for credit cards), the Fair Credit Billing Act, the CARD Act, and the Electronic Fund Transfer Act (for debit cards). These laws establish disclosure requirements, limit consumer liability for unauthorized transactions, and govern the dispute resolution process.

Real-time payment networks — the Clearing House's RTP network and the Federal Reserve's FedNow service, launched in 2023 — provide instant, final, 24/7 settlement as an alternative to ACH. They significantly raise the stakes for fraud controls because instant finality means fraudulent transfers cannot be reversed after they settle.

FinTech companies seeking banking capabilities have several charter options: full-service national or state bank charters; the industrial loan company charter, which allows commercial firms to own a depository institution without Federal Reserve holding company oversight — Walmart's 2005 ILC application was denied; and the OCC special purpose FinTech charter, which requires performing at least one core banking function and meeting capital, financial inclusion, and resolution planning requirements but does not require deposit-taking. The charter's legal status remained contested as of early 2026.

Bank partnerships remain the most common practical pathway for FinTech companies seeking banking capabilities. They provide access to deposit infrastructure, payment systems, and regulatory permissions without the obligations of a charter, but they create dependency on partner banks whose due diligence requirements, regulatory standing, and business decisions constrain what the FinTech can offer. The collapse of Synapse in 2024 illustrated the systemic risks that the bank partnership model can create when it scales without adequate regulatory oversight of the intermediary layer.

Chapter 6: WealthTech, Robo-Advisors, and the New Face of Financial Advice

Introduction

For most of American history, professional investment advice was a service for the wealthy. A financial adviser — someone who would sit across a desk, review your goals, construct a portfolio, and manage it over time — charged fees or commissions that made the relationship economically viable only for clients with substantial assets. The minimum investment thresholds at major advisory firms, the complexity of opening and funding an account, and the commission structures of broker-dealers all pointed toward the same conclusion: serious investment management was not designed for the average person.

Technology has been eroding that boundary for decades, but the emergence of robo-advisors accelerated the process significantly. A robo-advisor is an automated investment platform that uses algorithms to build and manage a diversified portfolio based on a client's stated goals, time horizon, and risk tolerance. The first standalone robo-advisors launched around 2010, in the immediate aftermath of the 2008 financial crisis — a moment when public trust in traditional financial advisers was low and technology-driven alternatives found a receptive audience. Betterment and Wealthfront were the pioneers; within a decade, virtually every major brokerage had launched its own automated advisory offering.

The rise of automated investment platforms is one dimension of a broader transformation in wealth management that goes by the name WealthTech. WealthTech encompasses robo-advisors, automated portfolio rebalancing tools, direct indexing platforms, account aggregation services, digital financial planning tools, and the algorithmic trading infrastructure that now underlies much of professional asset management. What unites these products is the application of technology to lower the cost, improve the accessibility, and increase the personalization of investment services that were previously available only to a fraction of the population.

This transformation raises questions that sit at the intersection of law, ethics, and technology — and that are worth taking seriously before they become invisible through familiarity. When an algorithm manages your retirement savings, who is responsible for the outcome? What duty does the platform owe you when its recommendation serves its own financial interests as well as yours? What does it mean to give financial advice when no human being is involved? And what happens when the platform's design — its notifications, its interface, its incentive structures — nudges you toward decisions that may not serve your long-term financial interests?

This chapter works through these questions by building from the regulatory foundation upward. It begins with the legal framework for investment advice — the distinction between investment advisers and broker-dealers and the standards of conduct each must meet. It then examines how that framework applies to robo-advisors and automated platforms. It covers the fiduciary duty concept in detail, because it is the central legal standard governing the advice relationship and because it will recur throughout the course. And it examines the ethical and regulatory concerns raised by gamification, algorithmic decision-making, and account aggregation — concerns that this chapter introduces and that Chapter 13 develops in the context of artificial intelligence more broadly.

The Regulatory Foundation: Advisers and Broker-Dealers

To understand how WealthTech is regulated, you need to understand the two legal categories that govern the provision of investment services in the United States: registered investment advisers and broker-dealers. These categories are not merely administrative distinctions — they reflect fundamentally different legal relationships with clients and impose fundamentally different obligations on the firms and individuals operating within them.

Registered Investment Advisers

A registered investment adviser, or RIA, is any person or firm that, for compensation, engages in the business of advising others on the value of securities or the advisability of investing in, purchasing, or selling securities. This definition, drawn from the Investment Advisers Act of 1940, is intentionally broad. It captures not just the human financial planner who meets with clients quarterly to review their portfolios, but also the automated platform that constructs and manages a portfolio based on an algorithm — and any firm whose primary business is managing other people's money for a fee.

Investment advisers with more than $110 million in assets under management must register with the Securities and Exchange Commission, or SEC. Smaller advisers register with state securities regulators. Registration requires filing Form ADV — a comprehensive disclosure document that describes the adviser's business, fees, conflicts of interest, disciplinary history, and the qualifications of its key personnel. Form ADV is publicly available through the SEC's Investment Adviser Public Disclosure database, giving prospective clients a standardized way to research and compare advisers.

The defining characteristic of a registered investment adviser is the fiduciary duty it owes to its clients. Fiduciary duty is a legal obligation to act in the best interest of another party — to put the client's interests ahead of your own. In the investment advice context, fiduciary duty has two components: the duty of loyalty, which requires the adviser to avoid conflicts of interest or, where conflicts cannot be avoided, to disclose them and manage them so that the client's interests are not subordinated; and the duty of care, which requires the adviser to provide advice that is suitable and well-reasoned given what the adviser knows about the client's goals, risk tolerance, and financial situation.

Fiduciary duty is a demanding standard. It means that an adviser who recommends a mutual fund that pays the adviser a higher commission than an equally suitable fund with a lower commission has violated the duty of loyalty — even if the recommended fund is not a bad investment in absolute terms. It means that an adviser who fails to update a client's portfolio in response to a material change in the client's circumstances has violated the duty of care. And it means that the legal obligation runs throughout the advisory relationship, not just at the moment a recommendation is made.

Broker-Dealers

A broker-dealer is a firm that is in the business of buying and selling securities — either as an agent on behalf of customers (broker) or as a principal for its own account (dealer). Broker-dealers execute trades, maintain customer accounts, underwrite securities offerings, and make markets in securities. They are regulated under the Securities Exchange Act of 1934 and must register with both the SEC and the Financial Industry Regulatory Authority, or FINRA — a self-regulatory organization that operates under SEC oversight and sets conduct standards, licensing requirements, and examination procedures for the broker-dealer industry.

The traditional standard of conduct for broker-dealers was the suitability standard: a broker-dealer making a recommendation had to have a reasonable basis to believe the recommendation was suitable for the customer, taking into account the customer's investment objectives, financial situation, and needs. The suitability standard is meaningfully lower than the fiduciary standard. A suitable recommendation is one that is appropriate for the client — it need not be the best option available, and the broker-dealer is not required to disclose all conflicts of interest or to place the client's interests ahead of its own financial interests in the way a fiduciary is.

The difference between fiduciary duty and suitability has practical consequences that are not subtle. A broker-dealer operating under a suitability standard could recommend a higher-commission product over a lower-commission product with identical characteristics without violating its legal obligations, so long as both products were suitable for the customer. The same recommendation by a fiduciary investment adviser would be a violation of the duty of loyalty.

Regulation Best Interest

The gap between the fiduciary standard applicable to investment advisers and the suitability standard applicable to broker-dealers created a persistent consumer protection problem: investors dealing with broker-dealers were exposed to conflicts of interest that investors dealing with RIAs were not. The problem was compounded by the fact that many consumers could not distinguish between the two categories — a broker-dealer and an investment adviser might offer what appeared to be similar services, but operate under very different legal obligations.

In 2019, the SEC adopted Regulation Best Interest, known as Reg BI, which raised the standard of conduct for broker-dealers without eliminating the distinction between broker-dealers and investment advisers. Under Reg BI, a broker-dealer making a recommendation of a securities transaction or investment strategy to a retail customer must act in that customer's best interest at the time of the recommendation — a standard explicitly higher than mere suitability, but still distinct from the full fiduciary duty that applies to investment advisers.

Reg BI has four core components. The disclosure obligation requires broker-dealers to provide retail customers with a standardized document — the Customer Relationship Summary, or Form CRS — that explains the nature of the firm's services, fees, conflicts of interest, and the legal standard of conduct applicable to the relationship. The care obligation requires the broker-dealer to exercise reasonable diligence, care, and skill in making recommendations and to put the retail customer's interests ahead of the broker-dealer's own financial interests. The conflict of interest obligation requires broker-dealers to establish and enforce policies and procedures to identify, disclose, and mitigate conflicts of interest. And the compliance obligation requires the firm to implement and maintain a compliance program designed to achieve adherence to Reg BI's requirements.

Reg BI was controversial when adopted. Consumer advocates argued it did not go far enough — that the best interest standard it established was insufficiently different from the suitability standard it replaced, and that it left in place conflicts of interest that a true fiduciary standard would prohibit. Industry groups argued it struck the right balance between investor protection and preserving broker-dealers' ability to serve retail customers profitably. The debate about whether Reg BI adequately protects retail investors from broker-dealer conflicts of interest has continued since its adoption and reflects the broader tension between investor protection and financial industry business models that runs through this field.

The table below summarizes the key distinctions between registered investment advisers and broker-dealers, including how robo-advisors fit within each category.

Feature	Registered Investment Adviser (RIA)	Broker-Dealer
Primary statute	Investment Advisers Act of 1940	Securities Exchange Act of 1934
Federal regulator	SEC (or state regulator for smaller firms)	SEC + FINRA (self-regulatory organization)

Standard of conduct	Fiduciary duty: must act in client's best interest at all times	Regulation Best Interest: must act in client's best interest at time of recommendation; ongoing duty is lower
Compensation model	Typically fee-based (flat fee, percentage of AUM, hourly)	Typically commission-based or transaction-fee-based
Conflicts of interest	Must be disclosed and managed; cannot be placed ahead of client interest	Must be disclosed and mitigated; conflict does not automatically disqualify recommendation
Ongoing obligations	Continuous duty of loyalty and care	Obligation primarily at point of recommendation
Robo-advisor registration	Must register as RIA; subject to full fiduciary standard	If operating as broker-dealer, subject to Reg BI

Robo-Advisors: How They Work and How They Are Regulated

A robo-advisor is an investment platform that uses algorithms to provide portfolio management services with minimal or no human involvement. The typical robo-advisor experience goes roughly like this: a new user completes an online questionnaire about their investment goals, time horizon, risk tolerance, and financial situation. The algorithm uses that information to assign the user to a model portfolio — typically a diversified mix of low-cost index funds or exchange-traded funds weighted across asset classes. The platform then automatically invests the user's contributions according to that model, rebalances the portfolio periodically to maintain the target allocation, and may apply tax-loss harvesting strategies to improve after-tax returns. The user can monitor their portfolio through a dashboard and adjust their settings at any time.

This service, which would have cost tens of thousands of dollars a year in minimum investment requirements and one percent or more of assets under management in annual fees at a traditional wealth management firm, is available from robo-advisors for annual fees as low as zero to 0.25 percent with no minimum balance at all. The cost reduction is a direct consequence of automation: the marginal cost of managing one additional account is near zero once the algorithm and infrastructure are built, regardless of the size of the account.

Regulatory Status of Robo-Advisors

Robo-advisors that provide portfolio management services for compensation are investment advisers under the Investment Advisers Act of 1940. The fact that the advice is delivered by an algorithm rather than a human being does not change the legal classification. A firm that charges fees to manage client portfolios — even entirely through automated means — is in the business of providing investment advice and must register with the SEC or the relevant state regulator depending on the size of its assets under management.

Betterment, Wealthfront, and other standalone robo-advisors are registered investment advisers subject to the full fiduciary standard. Their Form ADV filings are publicly available and disclose their fee structures, business models, and conflicts of interest. They are subject to SEC examination and enforcement, and their advertising and marketing materials are subject to the advertising rules applicable to all registered investment advisers.

The SEC has issued specific guidance on the application of the Investment Advisers Act to robo-advisors, acknowledging that automated platforms raise novel interpretive questions in several areas. The guidance addresses how robo-advisors should fulfill their fiduciary duty when the primary client interaction is through a digital questionnaire rather than a personal conversation, what disclosures are adequate when the algorithm's methodology is proprietary and not fully transparent to clients, and how robo-advisors should handle situations where a client's questionnaire responses suggest inappropriate risk taking or where the client's circumstances change between questionnaire updates.

The Questionnaire Problem

The initial questionnaire that robo-advisors use to assess client risk tolerance and investment goals is not merely a user experience feature — it is the primary mechanism through which the platform fulfills its duty of care. A registered investment adviser must have a reasonable basis for its recommendations given what it knows about the client. For a human adviser, that knowledge comes from conversations, documents, and an ongoing relationship. For a robo-advisor, it comes almost entirely from the questionnaire.

This creates a meaningful limitation. A questionnaire that asks about risk tolerance in the abstract may not adequately capture a client's actual financial situation — their existing assets, debts, income volatility, insurance coverage, tax situation, or the specific goals they have not thought to mention. A client who answers questions about investment volatility tolerance in the abstract may react very differently to an actual twenty percent decline in their portfolio value. And a client whose situation changes significantly — a job loss, a health crisis, an inheritance — may not update their questionnaire proactively, leaving the algorithm managing the portfolio based on stale information.

The SEC has encouraged robo-advisors to build processes for prompting clients to update their information and to consider whether clients' behavior — their withdrawal patterns, their response to market events, their account activity — should inform the platform's assessment of their actual risk tolerance. These prompts do not eliminate the fundamental limitation of questionnaire-based advice, but they represent an attempt to make the robo-advisor's fulfillment of its duty of care more dynamic and less mechanical.

Hybrid Models

The sharp distinction between pure robo-advisors and human financial advisers has blurred considerably since the first platforms launched. Most major robo-advisors now offer some form of access to human advisers — either as a premium tier or as an option for clients with more complex needs. Betterment Premium offers human adviser access to clients above a certain asset threshold. Vanguard's Personal Advisor Services pairs algorithmic portfolio management with human financial planning conversations. Schwab Intelligent Portfolios Premium combines automated investing with unlimited access to a certified financial planner.

Traditional brokerage firms and wealth management companies have moved in the opposite direction, building or acquiring robo-advisory capabilities to serve client segments that their traditional human adviser model could not serve cost-effectively. Charles Schwab, Fidelity, Merrill Lynch, and virtually every major wealth management firm now offers automated investment services alongside their traditional offerings.

The practical result is a spectrum rather than a binary: from fully automated robo-advisors at one end, through hybrid platforms that blend automation with human touchpoints, to fully human advisory relationships at the other. The regulatory framework applies fiduciary duty to any firm that registers as an investment adviser, regardless of where on that spectrum it sits. The interesting regulatory questions arise at the boundaries — what obligations does a hybrid platform have when its human advisers give recommendations that conflict with the algorithm's model portfolio, for example, or when a client interacts primarily with the algorithm but occasionally consults a human adviser on specific questions?

Fiduciary Duty in the Algorithmic Context

Fiduciary duty was developed as a legal concept in the context of human relationships — trustees managing property for beneficiaries, lawyers representing clients, financial advisers managing portfolios. Its application to algorithmic decision-making raises questions that the traditional doctrine does not cleanly answer.

Can an Algorithm Have a Fiduciary Duty?

The technical answer is no — a legal duty is an obligation imposed on a person or entity, not on a software program. The fiduciary duty in the robo-advisor context belongs to the registered investment adviser — the firm — not to the algorithm it runs. The firm is responsible for ensuring that its algorithm is designed and operated in a way that fulfills the fiduciary obligations the firm owes to its clients. If the algorithm systematically produces recommendations that are not in clients' best interests — because it is designed to maximize the platform's revenue, because it fails to account for relevant client information, or because it is simply poorly built — the firm, not the algorithm, is liable.

This distinction is important and has practical implications. A robo-advisor that designs its algorithm to recommend higher-fee products when lower-fee alternatives would serve clients equally well has a conflict of interest that the fiduciary standard requires it to disclose and manage. The fact that the biased recommendation comes from an algorithm rather than a human adviser does not reduce the firm's legal obligation — it may, if anything, amplify the harm, because an algorithmic bias operates at scale, affecting every client whose portfolio the algorithm manages simultaneously.

Transparency and the Black Box Problem

One of the most significant challenges in applying fiduciary duty to algorithmic advisers is transparency. The duty of care requires an adviser to provide sound recommendations based on its knowledge of the client and its understanding of the investment landscape. But what does that requirement mean when the recommendation comes from an algorithm whose internal logic is not fully visible even to the firm that built it?

Most robo-advisor algorithms are not genuinely opaque — they are rule-based systems that follow defined decision trees and optimization frameworks that the firm's designers can explain in general terms. But as the sophistication of the algorithms increases — as machine learning techniques are applied to portfolio optimization, tax-loss harvesting, and client risk assessment — the recommendations they produce may become harder to explain and audit. A machine learning model that has identified a pattern in historical data may generate a recommendation that no human reviewer can fully trace to a comprehensible rationale.

The SEC has signaled that the transparency of algorithmic recommendations is a fiduciary concern. Clients are entitled to understand the basis for the advice they receive, and a firm that cannot explain why its algorithm recommended a particular portfolio allocation to a particular client may struggle to demonstrate that it has fulfilled its duty of care. This tension between algorithmic sophistication and explainability is one of the themes that will recur in Chapter 13 when we examine artificial intelligence in FinTech more broadly.

Conflicts of Interest in Automated Platforms

Conflicts of interest in the robo-advisor context are real and varied. A platform that offers its own proprietary funds alongside third-party funds in its model portfolios has a financial incentive to allocate client assets to the proprietary products, which generate management fee revenue for the platform in addition to the advisory fee the client pays. A platform that earns revenue from cash sweep accounts — holding uninvested client cash in accounts that pay below-market interest rates while earning the spread — has an incentive to maintain higher cash balances in client portfolios than the client's investment strategy would otherwise require.

The SEC has identified both of these conflicts in enforcement actions and guidance. Schwab Intelligent Portfolios, for example, was the subject of an SEC enforcement action in 2022 resolving allegations that the platform's cash allocation practices were driven in part by Schwab's revenue interests rather than solely by clients' investment needs — and that the platform had not adequately disclosed this conflict. Schwab agreed to pay $187 million to settle the charges without admitting or denying the SEC's findings. The case illustrated that even well-capitalized, established financial institutions can have conflicts of interest embedded in their robo-advisor products — and that the fiduciary standard requires those conflicts to be disclosed and managed, not simply buried in fine print.

The conflict of interest framework is a powerful analytical tool when evaluating WealthTech companies. Any time an automated investment platform earns revenue in a way that could influence its recommendations — through proprietary fund allocations, cash sweep arrangements, payment for order flow, or premium service tiers — the fiduciary question is whether the conflict is disclosed and whether the recommendations can still be characterized as acting in the client's best interest.

Robinhood and the Gamification of Investing

No single company better illustrates both the democratizing potential and the ethical risks of WealthTech than Robinhood. Founded in 2013, Robinhood pioneered commission-free stock trading through a mobile-first interface designed to make investing as simple and engaging as a consumer app. It eliminated the minimum balance requirements and per-trade commissions that had long made active stock trading economically viable only for investors with substantial assets. Within a few years, it had attracted millions of users who had never previously owned individual stocks.

The product choices Robinhood made in designing that experience were not neutral. The app deployed features drawn from behavioral psychology and consumer app design — streak notifications congratulating users on consecutive days of activity, confetti animations celebrating trades, a simplified interface that reduced investing to a series of swipes and taps, and prominently displayed options trading features that were readily accessible to investors with limited experience. These design choices were not accidental. They were deliberate attempts to maximize user engagement — and they worked.

The GameStop Episode

In late January 2021, a group of retail investors coordinating on the Reddit forum WallStreetBets drove the price of GameStop — a struggling brick-and-mortar video game retailer — from roughly $20 to nearly $500 per share in a matter of days. The mechanics involved short squeeze dynamics: institutional investors had heavily shorted GameStop stock, expecting it to decline; retail investors buying en masse drove the price up, forcing short sellers to cover their positions at enormous losses by purchasing the stock at inflated prices, which drove the price even higher.

The GameStop episode exposed structural tensions in the market that regulators and commentators are still working through. But for Robinhood specifically, the defining moment came on January 28, 2021, when the platform abruptly restricted customers from purchasing GameStop and several other heavily traded stocks, allowing only sell orders. Robinhood's explanation was that the Depository Trust and Clearing Corporation had increased its deposit requirements for clearing volatile securities, and that Robinhood needed to limit purchase orders to meet those requirements without exhausting its capital.

The trading restriction triggered immediate and intense backlash. Retail investors who had been unable to buy stock during a critical period of price appreciation argued they had been prevented from participating in gains that institutional investors continued to access. Congressional hearings were called. Regulators launched inquiries. Bankman-Fried and other crypto figures offered Robinhood's situation as an argument for blockchain-based settlement that would eliminate clearing house deposit requirements. Robinhood's CEO testified before Congress to explain the company's decision.

The regulatory and legal aftermath was significant. FINRA fined Robinhood $70 million in 2021 — the largest financial penalty in FINRA's history at the time — for a range of violations including providing false and misleading information to customers, approving customers for options trading without adequate due diligence, and displaying misleading account information that caused some customers to believe they had negative balances. The Massachusetts Securities Division sought to revoke Robinhood's broker-dealer license in the state, alleging that its gamification practices were inconsistent with its obligations to customers. The Massachusetts proceeding was ultimately unsuccessful, but it focused public attention on whether gamification in investment platforms crosses the line from user-friendly design to manipulation.

The Gamification Problem

Gamification refers to the application of game design elements — rewards, streaks, challenges, progress indicators, visual reinforcement — to non-game contexts to drive engagement and behavior. In consumer technology, gamification is ubiquitous and largely benign: fitness apps award badges for reaching step goals; language learning apps track streaks; social media platforms display engagement metrics that encourage more posting. The behavioral dynamics are well understood and the harms, while real in some cases, are generally modest.

In the context of investing, gamification raises concerns of a different magnitude. Investment decisions have long-term financial consequences that cannot be undone with a swipe. The psychological mechanisms that gamification exploits — the desire for rewards, the aversion to breaking streaks, the social comparison that leaderboards and visible activity create — can drive investors toward more frequent trading, higher-risk positions, and shorter time horizons than their actual financial interests recommend. A retail investor who checks their portfolio daily and receives push notifications about market movements is likely to trade more than one who reviews their investments quarterly — and frequent trading generally produces worse long-term outcomes for retail investors due to transaction costs, taxes on realized gains, and the difficulty of timing markets accurately.

The regulatory question is whether Robinhood's design choices violated its obligations to customers. Under Reg BI, a broker-dealer must act in the retail customer's best interest at the time of a recommendation. The harder question is whether the platform's design — the features that encouraged more frequent trading, the prominent display of options trading to inexperienced investors, the congratulatory notifications — constituted implicit recommendations that triggered Reg BI obligations, or whether they were merely a user interface that customers could use or ignore as they chose.

The SEC has explored whether the use of digital engagement practices — gamification features, behavioral nudges, and AI-driven personalization — by broker-dealers and investment advisers creates obligations under Reg BI and the fiduciary standard. This remains an active regulatory question. What is clear is that the design of an investment platform is not ethically or legally neutral. Choices about what information to display prominently, what actions to make easy, and what behavioral cues to embed in the interface are choices that affect how customers behave — and firms that profit from customers' trading activity have financial incentives that may not align with customers' long-term interests.

Options Trading and Duty of Care

One of the most specific regulatory concerns about Robinhood's practices was its approval of customers for options trading. Options are complex derivative instruments that can generate large gains but can also result in losses exceeding the investor's initial investment. A call option gives the buyer the right to purchase a stock at a specified price before a specified date; if the stock does not reach that price, the option expires worthless. A put option gives the buyer the right to sell a stock at a specified price; if the stock price does not fall below that level, the option expires worthless. More complex options strategies — spreads, straddles, combinations — involve multiple positions and require sophisticated understanding of risk management.

For broker-dealers, the approval of customers for options trading has historically been subject to specific due diligence requirements: the broker-dealer must have a reasonable basis to believe the customer has the knowledge and experience to understand the risks of options trading. Robinhood's approval process, according to FINRA's findings, was inadequate — it approved customers for options trading based on self-reported information without adequate verification, and it did not apply the scrutiny that the complexity and risk of the products warranted. Several Robinhood customers sustained significant losses in options positions they did not fully understand, and in at least one documented case a young investor who misunderstood a displayed balance took his own life after believing he had lost far more than he had.

This tragic outcome is an extreme illustration of a broader principle: the design choices embedded in a financial platform have real consequences for real people. The obligation of financial services firms to understand who their customers are, what they are capable of handling, and what products are appropriate for them is not a bureaucratic formality — it is a substantive protection that exists because the consequences of getting it wrong are serious.

Account Aggregation and Open Banking in Wealth Management

One of the most significant developments in WealthTech is the emergence of account aggregation platforms — services that pull together information from multiple financial accounts into a single dashboard, giving users a comprehensive view of their financial lives.

Mint, Personal Capital (now Empower), and similar platforms allow users to connect their bank accounts, investment accounts, credit cards, and loans, and see their complete financial picture in one place. For users, the value is obvious: instead of logging into four different apps and maintaining a mental model of their overall financial situation, they see everything in one view, with automatic categorization, spending analysis, and net worth tracking.

Account aggregation creates value by solving a real problem, but it does so in ways that raise significant privacy, security, and regulatory questions. To aggregate account information, these platforms need access to data held by the user's various financial institutions. The traditional mechanism for obtaining that access was screen scraping — the platform would use the user's login credentials to log into each financial account and extract the data, as if it were the user. Screen scraping is technically functional but operationally risky: it requires users to share their passwords with a third party, gives the aggregator broader access to accounts than is strictly necessary for aggregation, and creates liability questions if the aggregator's access is misused.

Open banking — the provision of application programming interfaces, or APIs, by financial institutions that allow authorized third parties to access account data without requiring password sharing — is the more technically sound alternative. When a bank provides an API for data access, the aggregator can obtain the specific data the user has authorized without accessing the full account, and the authorization can be revoked at any time without changing the user's password. The CFPB's rulemaking under Section 1033 of the Dodd-Frank Act, finalized in 2024, establishes standards for open banking data sharing in the United States, requiring financial institutions to provide consumers and authorized third parties with machine-readable access to consumer financial data. This rule has significant implications for the WealthTech ecosystem — it creates a legal right for consumers to share their financial data with account aggregators and investment platforms, and it establishes the consumer protection standards that govern how that data must be handled.

For WealthTech platforms that use aggregated account data to provide financial planning or investment advice, the regulatory questions are layered. The aggregator itself may or may not be providing investment advice — if it is merely displaying account information without making recommendations, it may not be an investment adviser. But a platform that uses aggregated data to generate personalized portfolio recommendations, identify investment opportunities, or provide holistic financial planning is almost certainly providing investment advice and must register accordingly. The combination of comprehensive personal financial data with algorithmic advice generation is precisely the kind of product that regulators are watching closely as the boundaries of the Investment Advisers Act are tested by new technology.

The Ethical Dimensions of Algorithmic Financial Advice

Chapter 6 closes the Module 3 content, but it opens a set of ethical questions that will run through the remainder of the course. The brief treatment of these questions here is intentional — the full ethical framework is developed in Chapters 7 and 8 on data ethics and privacy law, and the AI-specific dimensions are addressed in Chapter 13. But introducing the ethical concerns in the WealthTech context grounds them in something concrete before the more theoretical treatment.

Access and Equity

The democratization narrative around robo-advisors is genuinely compelling: technology has made professional-quality portfolio management available to people who previously could not access it. A 22-year-old starting her career with $1,000 in savings can now get the same diversified, rebalanced, tax-efficient portfolio management that a wealthy investor with $1 million would have received from a traditional adviser. That is a real and meaningful expansion of financial opportunity.

But the democratization is not complete. The financial benefits of automated investing — compounding returns over long time horizons, tax-loss harvesting, diversified exposure to capital markets — accrue disproportionately to people who can afford to invest and leave their investments alone. Someone living paycheck to paycheck, facing irregular income, or carrying high-interest debt is not well served by a robo-advisor that optimizes for long-term wealth accumulation. The products that FinTech has been slowest to develop for underserved populations are often the ones that matter most: accessible credit at fair rates, tools for building emergency savings, insurance products for irregular income workers. Robo-advisors serve people who already have surplus income to invest; they do not yet solve the financial access problems that affect the broadest range of Americans.

Algorithmic Accountability

When a human financial adviser makes a bad recommendation, there is a person to hold accountable — someone who can explain their reasoning, acknowledge their error, and be subject to professional discipline or legal liability. When an algorithm makes a bad recommendation, accountability is more diffuse. The firm is legally responsible, but identifying what went wrong inside a complex algorithm and demonstrating that the error was the result of negligence rather than an unfortunate but reasonable outcome requires technical and legal analysis that is expensive and time-consuming.

This accountability gap has practical consequences for clients who receive bad algorithmic advice. It also has regulatory consequences: a supervisory framework designed to hold human advisers accountable through their professional licenses and conduct records needs adaptation to hold algorithmic advisers accountable in comparable ways. The SEC has been developing its thinking on algorithmic accountability, and this is a rapidly evolving area of regulatory concern that Chapter 13 addresses in more depth.

The Data Dimension

Investment platforms collect substantial amounts of personal financial data — transaction histories, portfolio performance, spending patterns, income flows, risk tolerance responses, and behavioral data about how users interact with the platform. This data is valuable far beyond the investment advice context. It reveals consumer behavior, financial health, lifestyle patterns, and life events in ways that have commercial value to marketers, lenders, insurers, and a range of other parties.

The ethical question of what WealthTech firms can and should do with the financial data they collect will be examined in detail in Chapter 7. For now, the key point is that the investment platform relationship involves an asymmetry of information and power: the platform knows a great deal about the user, and the user often has limited visibility into how that data is being used, shared, or monetized. The fiduciary standard's requirement that investment advisers put clients' interests first provides some protection against data misuse within the advisory relationship — a robo-advisor that sells client portfolio data to a competitor, for example, would likely violate its duty of loyalty. But the fiduciary standard does not cover all the ways financial data can be used, and the broader regulatory framework for data privacy — addressed in Chapter 8 — fills some but not all of the gaps.

Looking Ahead: WealthTech and Artificial Intelligence

Chapter 6 has introduced the foundational regulatory framework for WealthTech and identified the key ethical concerns that automated investment platforms raise. Several of those concerns — algorithmic transparency, conflicts of interest embedded in recommendation systems, the accountability gap in algorithmic decision-making — are versions of broader questions about artificial intelligence in financial services that Chapter 13 addresses directly.

The robo-advisor of 2010 was a rule-based algorithm: a relatively simple set of decision rules about asset allocation, rebalancing triggers, and tax-loss harvesting parameters. The investment platforms of the mid-2020s are increasingly incorporating machine learning techniques that adapt to market conditions, client behavior, and economic data in ways that are harder to explain and audit than the original rule-based systems. As these systems become more sophisticated, the regulatory questions become correspondingly more complex: how do you audit a machine learning model for bias or conflicts of interest? How do you explain a recommendation that emerges from a system that has identified patterns in thousands of variables simultaneously? How do you design governance structures that provide meaningful oversight of systems that operate at a scale and speed that exceeds human review capacity?

These questions do not have complete answers today. They are the frontier of both regulatory development and industry practice. Students who understand the foundational concepts covered in this chapter — fiduciary duty, conflicts of interest, the standards applicable to investment advisers and broker-dealers, and the ethical concerns raised by gamification and algorithmic advice — will be well equipped to engage with those more complex questions when they encounter them in Chapter 13 and in professional practice.

Key Takeaways

WealthTech is the application of technology to investment management, financial planning, and related services. Robo-advisors — automated investment platforms that use algorithms to construct and manage portfolios — are the most prominent WealthTech product category and have significantly lowered the cost and minimum investment threshold for professional portfolio management.

The two primary legal categories governing investment services are registered investment advisers and broker-dealers. Registered investment advisers are subject to the Investment Advisers Act of 1940 and owe clients a fiduciary duty — a legal obligation to act in the client's best interest, comprising a duty of loyalty and a duty of care. Broker-dealers are regulated under the Securities Exchange Act of 1934 and FINRA rules, and are subject to Regulation Best Interest, adopted in 2019, which requires acting in a retail customer's best interest at the time of a recommendation but does not impose the same continuous fiduciary obligation as the investment adviser standard.

Robo-advisors that manage client portfolios for compensation are investment advisers under the Investment Advisers Act and must register with the SEC or the relevant state regulator. They are subject to the full fiduciary standard, including obligations around conflict of interest disclosure and the duty of care in algorithmic recommendation design.

Fiduciary duty in the algorithmic context attaches to the firm, not the algorithm. A robo-advisor that embeds conflicts of interest in its algorithm — by systematically recommending higher-fee proprietary products, for example — violates its fiduciary obligation even though the biased recommendation comes from a software system. Transparency about algorithmic methodology and robust conflict of interest management are both requirements of the fiduciary standard as applied to automated platforms.

Robinhood's experience illustrates both the potential and the risks of applying consumer technology design principles to investment platforms. Gamification features — confetti animations, streak notifications, simplified interfaces for complex products — may drive user engagement in ways that conflict with users' long-term financial interests. FINRA's $70 million fine against Robinhood addressed specific violations including inadequate due diligence for options trading approval and misleading account displays. The ethical and regulatory questions raised by gamification remain active areas of regulatory development.

Account aggregation platforms — services that consolidate financial account information from multiple institutions — raise privacy, security, and regulatory questions that are addressed more fully in Chapters 7 and 8. The CFPB's Section 1033 open banking rule, finalized in 2024, establishes standards for consumer-directed data sharing that have significant implications for the WealthTech ecosystem.

The ethical concerns introduced in this chapter — algorithmic accountability, the limits of democratization, conflicts embedded in recommendation systems, and the data dimension of investment platforms — will be developed more fully in the chapters that follow. Chapter 13 addresses artificial intelligence in financial services, including the machine learning techniques increasingly applied to investment management, the algorithmic bias concerns they raise, and the emerging regulatory response.

Chapter 7: Data Ethics — Privacy, Consent, and the Value of Consumer Data

Introduction

In 1999, a children's toy retailer called Toysmart collected personal information from its customers — names, addresses, family profiles, and children's ages and wish lists. On its website, Toysmart displayed the TRUSTe privacy seal, a widely recognized signal that a website had committed to responsible privacy practices. The company's privacy policy was explicit: personal information would never be sold or shared with third parties.

The following year, Toysmart filed for bankruptcy. Its most valuable remaining asset was that customer database — the same data it had promised never to sell. Toysmart's management, facing creditors owed millions of dollars, proposed to sell the database as part of the bankruptcy proceedings. The Federal Trade Commission intervened. Privacy advocates objected. Customers who had trusted the company with their children's information had no say in the outcome.

What should Toysmart have done? That question does not have an obvious answer, and this chapter gives you the tools to work through it rigorously. The ethical frameworks taught here — applied to the facts of the Toysmart case — illustrate how data ethics analysis works in practice. Chapter 8 returns to the case with the legal frameworks that complete the picture.

Data ethics is not an abstract academic exercise. It is a practical discipline that equips professionals to recognize ethical problems, reason through competing interests, and arrive at positions they can defend with evidence and argument. FinTech is a field saturated with exactly these problems — firms that collect sensitive financial data at scale, that monetize that data in ways consumers may not anticipate, and that face recurring tension between commercial interests and the privacy expectations of the people whose data they hold. Professionals who can only ask what is legally permitted are poorly equipped for this environment. This chapter builds the capacity to ask what is right.

What Is Ethics?

Ethics is a rational examination of people's moral beliefs and actions. It is a branch of philosophy concerned with determining which activities are good and which are bad, which actions are right and which are wrong, and how people ought to behave toward one another. It is a field of study more than 2,000 years old, with roots in ancient Greek philosophy and contributions from virtually every major intellectual tradition since.

Ethics is frequently confused with three related but distinct concepts: legality, morality, and personal preference. Legality refers to what the law permits or prohibits. Ethics refers to what is right or wrong independent of what is legal — the two often align, but they do not always. A practice can be entirely legal and still be ethically wrong. A practice can be ethically required even when no law mandates it. Morality typically refers to a particular set of values held by an individual or community, often rooted in culture, religion, or tradition. Ethics, as a discipline, involves reasoning about moral questions rather than simply applying a fixed moral code. Personal preference — liking or disliking something — is not the same as an ethical judgment about whether it is right or wrong.

The distinction between ethics and legality is not merely semantic. It defines the scope of professional responsibility. A FinTech company that asks only what is legally permitted has not asked the right question. The legally permitted action is the floor, not the ceiling. Professionals who treat legal compliance as the complete answer to ethical questions will make decisions that harm real people while technically following the rules — and they will be unprepared when those decisions are later scrutinized.

The Philosophical Foundations of Ethical Reasoning

Ethical reasoning draws on several distinct philosophical traditions. These are not competing belief systems — they are different analytical lenses, each illuminating aspects of a situation that the others may miss. Skilled ethical reasoners do not pick one framework and ignore the rest. They use multiple frameworks to build a more complete picture of what is at stake.

Social Contract Theory

Social contract theory holds that people agree, implicitly or explicitly, to accept rules and constraints on their behavior in exchange for the benefits of living in an organized society. Without such an agreement — without rules and a means of enforcing them — life would be, as Thomas Hobbes described it, solitary, poor, nasty, brutish, and short. Hobbes called this condition the state of nature: the condition of human life absent social organization, shared rules, and enforcement mechanisms.

Social contract theory grounds ethical obligations in the idea of mutual agreement and reasonable expectations. When a company makes promises to its customers — about how their data will be used, what services they will receive, how disputes will be handled — it is entering into a form of social contract. Those promises create obligations that exist independently of whether the law specifically enforces them. Breaking an explicit promise, even when legally defensible, violates the reasonable expectations on which the relationship was based.

In the FinTech context, social contract reasoning is particularly relevant because the entire customer relationship depends on trust. Consumers share sensitive financial information with FinTech platforms because they trust, based on explicit and implicit promises, that the information will be used appropriately. When that trust is violated, the harm extends beyond the individual transaction — it damages the broader social contract between technology companies and the public that makes data-driven services possible.

Utilitarianism

Utilitarianism is a consequentialist theory of ethics. It evaluates the rightness or wrongness of an action based entirely on its consequences: the morally correct action is the one that produces the greatest good for the greatest number. Associated with philosophers Jeremy Bentham and John Stuart Mill, utilitarianism does not judge actions by fixed rules or by the character of the actor — it judges them by their outcomes.

Calling utilitarianism a consequentialist theory is precise and important. It means that the same action can be right in one context and wrong in another, depending on what consequences it produces. It means that good intentions are not sufficient — outcomes matter. And it means that trade-offs are a legitimate part of ethical analysis: if an action harms some people but benefits many more, utilitarianism provides a framework for weighing those interests against each other.

In data ethics, utilitarian reasoning frequently appears in arguments about data aggregation and analytics. A credit bureau might argue that compiling detailed credit histories on millions of consumers serves the public good by making lending markets more efficient and expanding credit access — even though the compilation involves collecting and using personal information that individuals might prefer to keep private. A fraud detection system that monitors every transaction to identify suspicious patterns might be justified on utilitarian grounds even if some consumers find it intrusive. These arguments are not necessarily wrong, but they require honest accounting of both the benefits and the harms — including harms to the individuals whose privacy interests are overridden in the name of aggregate benefit.

Deontological Ethics

Deontological ethics, most closely associated with Immanuel Kant, evaluates actions based on their consistency with universal moral duties and principles rather than their consequences. The central Kantian principle — the categorical imperative — holds that one should act only according to rules one could will to become universal laws, and that one should always treat people as ends in themselves, never merely as means.

The categorical imperative provides a distinctive test for ethical conduct: could you will that everyone in a similar situation act the same way? If a data practice could not be universally adopted without contradiction or harm — if the practice depends on most people not knowing about it, for example, or on a general expectation of privacy that the practice itself undermines — then it fails the categorical imperative test.

The second component — treating people as ends in themselves — has direct application to data monetization. When a FinTech platform sells consumer data to advertisers without meaningful disclosure or consent, it is using consumers as instruments of its own commercial objectives. The consumers are not participants in a mutually beneficial exchange; they are resources being converted into revenue without their knowledge. Deontological ethics identifies this as categorically wrong, regardless of the aggregate economic benefit it produces.

Consider the practice of price personalization — adjusting the loan terms or product prices shown to individual consumers based on inferred willingness to pay, derived from behavioral data the consumer did not know was being analyzed. A utilitarian might argue this is acceptable if it expands access to credit or maximizes market efficiency. A deontologist would argue that using private behavioral data to extract the maximum payment from each individual treats consumers as targets to be optimized against rather than people whose interests deserve honest dealing.

Virtue Ethics

Virtue ethics, rooted in Aristotle, shifts the focus from the action and its consequences to the character of the person or organization performing it. The central question is not what does the rule require, or what outcome does this produce, but what would a person — or company — of genuine integrity and good character do in this situation?

In a professional context, virtue ethics maps naturally onto professional standards and organizational culture. A data scientist who notices that a credit scoring model she is building incorporates proxy variables correlated with race has a choice: she can flag the problem and advocate for its correction, or she can stay quiet and ship the model on schedule. Virtue ethics asks what an honest professional of good character would do — not merely what her employment contract requires.

Virtue ethics also captures something important about the difference between genuine ethical practice and compliance theater. A FinTech company can design its consent flows and privacy disclosures to technically satisfy legal requirements while being written in ways that most consumers will not read, presented in formats designed to minimize engagement, and framed to obscure the actual scope of what is being consented to. Such a company is complying with the letter of the law while violating any honest standard of transparent dealing. Virtue ethics identifies the gap between formal compliance and genuine integrity — and asks which one the organization is actually practicing.

Stakeholder Theory

Stakeholder theory is an essential analytical tool for applied ethics in business settings. The core idea is that significant business decisions affect a range of parties beyond the company itself, and that ethical analysis requires identifying all affected parties — stakeholders — and considering the impact on each.

In data ethics, stakeholder analysis is particularly important because data collection and use affect parties who are often invisible in the decision-making process. When a FinTech platform decides to sell consumer transaction data to a health insurance company, the affected stakeholders include the consumers whose data is sold, the health insurance company and its policyholders, employers who may use the insurance data in employment decisions, and the broader public whose trust in financial privacy norms is implicated. A business decision that looks straightforward from the perspective of the firm's revenue interests looks very different when all affected stakeholders are visible.

Stakeholder theory does not resolve conflicts between competing interests — it maps them. Knowing that a decision harms some stakeholders while benefiting others does not automatically tell you what to do. That judgment requires applying the other ethical frameworks to weigh whose interests should prevail and why. But stakeholder analysis prevents the analytical error of treating only the firm's interests as real, which is the starting point for a great deal of unethical behavior in the data economy.

The table below summarizes the major ethical frameworks and their application to data privacy scenarios.

Framework	Core Question	How It Evaluates an Action	Strength	Limitation in Data Ethics
Utilitarianism (Consequen-tialist)	What produces the greatest good for the greatest number?	Evaluates consequences: does the action produce more benefit than harm overall?	Practical; weighs real-world outcomes; accommodates trade-offs	Can justify harm to individuals if aggregate benefit is large enough; difficult to measure and compare harms across stakeholders

Deontological Ethics (Kantian)	Does this action respect people as ends in themselves, not merely as means?	Evaluates duties and rules: is the action consistent with universal moral principles regardless of outcome?	Protects individual rights regardless of majority preferences; provides clear prohibitions	Can be rigid; may prohibit actions with beneficial outcomes; difficult to resolve conflicts between duties
Virtue Ethics	What would a person of good character do?	Evaluates the character and intentions of the actor, not just the action or its consequences	Focuses on integrity and professional responsibilit y; resonates with professional codes of ethics	Less prescriptive; depends on cultural definitions of virtue; harder to operationali ze in policy
Social Contract Theory	Does this action honor the implicit agreement between parties?	Evaluates whether an action is consistent with what people would agree to under fair conditions	Grounds ethics in mutual obligation and reasonable expectations; aligns with consent frameworks	Social contracts are implicit and contested; what people 'would agree to' is debatable
Stakeholder Theory	Who is affected and how?	Identifies all parties with interests at stake and evaluates the impact on each	Comprehens ive; prevents harm to overlooked parties; maps conflicts between stakeholders	Does not resolve conflicts between stakeholder interests; requires judgment about whose interests prevail

Privacy as an Ethical Value

Privacy is not simply a legal concept — it is an ethical value grounded in the idea that people have a fundamental interest in controlling information about themselves and in being free from unauthorized observation. Understanding privacy as an ethical value, rather than merely a compliance obligation, shapes how FinTech professionals think about what they build and how they use the data they collect.

What Privacy Protects

Privacy protects a cluster of related interests. It protects autonomy — the ability to make decisions about one's own life without being observed, judged, or constrained by the knowledge that others are watching. It protects dignity — the sense that one's personal information and relationships are not simply available for commercial exploitation. It protects safety — the ability to share information in one context without that information being used against one in a different context. And it protects trust — the ability to form relationships based on the reasonable expectation that shared information will be treated with discretion.

Financial information is among the most sensitive personal information that exists. A person's income, debt load, spending patterns, investment holdings, and transaction history reveal their life circumstances, vulnerabilities, and behaviors in ways that can be exploited by lenders adjusting credit terms, insurers pricing policies, employers making hiring decisions, or bad actors committing identity fraud. The sensitivity of financial data is precisely why the relationship between a financial institution and its customers has historically been governed by strong confidentiality norms, even when those norms were not explicitly required by law.

Public vs. Private Information

Not all information about a person is private. A public record contains information about an incident or action reported to a government agency for the purpose of informing the public. Court records, property tax records, business filings, and government contracts are public records because the public interest in transparency outweighs the individual's privacy interest with respect to those specific matters.

The distinction between public and private information has become significantly more complex in the digital age. Information that was technically public — a court record, a newspaper article, a social media post — but practically obscure because accessing it required physical effort has become a different kind of public when it is indexed, aggregated, and made instantly searchable online. The legal status of information does not fully capture its ethical status: information that was private in practice, even if technically public in law, raises ethical concerns when technology changes its practical accessibility.

This aggregation problem is central to contemporary data ethics. Data analytics platforms routinely combine information from multiple sources — some explicitly public, some provided directly by consumers, some obtained from third-party data brokers — to build detailed profiles of individuals. The ethical question is not merely whether each individual piece of information was public or private at the time it was collected. It is whether the assembled profile represents the kind of comprehensive surveillance that the individual could not reasonably have understood herself to be consenting to when she shared each piece separately.

Invisible Information Gathering

One of the most significant privacy concerns in the digital economy is invisible information gathering — the collection of personal data through processes the individual is unaware of or does not understand. When a person visits a website, their IP address, the links they clicked, the time spent on each page, and a range of other behavioral signals are automatically recorded — without any explicit notice or request for consent. This automatic recording is a straightforward example of invisible information gathering.

Invisible information gathering is pervasive in FinTech. Mobile payment apps collect location data that reveals where and when purchases are made. Trading platforms track exactly how users interact with their interface — which securities they view, how long they examine each, what information they access before placing an order. Credit scoring models incorporate behavioral signals that consumers do not know are being evaluated. Fraud detection systems build behavioral profiles of normal account activity without disclosing to customers what data points constitute the profile or how deviations trigger review.

The ethical concern with invisible information gathering is not that it occurs — some of it serves legitimate purposes and benefits consumers — but that it happens without meaningful transparency or consent. When consumers cannot know what data is being collected about them, they cannot make informed decisions about which services to use, cannot understand the basis for decisions that affect them, and cannot hold companies accountable for misuse. The gap between what consumers believe is happening with their data and what is actually happening is a structural problem in the data economy that FinTech firms both benefit from and contribute to.

Reasonable Expectation of Privacy

The U.S. Supreme Court established the legal concept of reasonable expectation of privacy in Katz v. United States in 1967. The case involved FBI agents who had attached a listening device to a public telephone booth without obtaining a warrant, intercepting the defendant's phone calls. The Court held that the government needs a court order to intrude upon communications or activities in which a person has a reasonable expectation of privacy — that what a person seeks to preserve as private, even in an area accessible to the public, may be constitutionally protected.

Katz established that the Fourth Amendment protects people, not places. The relevant question is not whether the information is in a physically private location but whether the person had a reasonable expectation that it would remain private. A person making a phone call in a public booth has a reasonable expectation that her conversation is private, even though she is in a public space. The principle has been extended and adapted to new technologies in subsequent decisions, with significant implications for data collection in the digital context.

As an ethical standard — not just a legal one — reasonable expectation of privacy asks what uses of personal data a reasonable person would have anticipated and consented to at the time of disclosure. A consumer who provides a name and payment information to complete a purchase has a reasonable expectation that the data will be used to process the transaction and, perhaps, to improve the service she receives. She does not have a reasonable expectation that her purchase history will be sold to a data broker, incorporated into a health risk model, or used to adjust the prices she sees on future visits. The distance between what was reasonably expected and what actually happens is a reliable measure of ethical risk.

Technology, Data, and the Erosion of Privacy Expectations

The technologies that define modern FinTech have created new capabilities for data collection that existing privacy frameworks did not anticipate. Radio-frequency identification, or RFID, uses wireless transmission to identify and track objects and people. An RFID chip is a wireless transmitter — it emits a signal readable by a compatible reader without physical contact and often without the awareness of the person carrying the device. RFID technology is embedded in contactless payment cards, passports, transit cards, and employee badges, enabling fast and frictionless transactions while also creating the potential for passive tracking of locations and activities.

The broader landscape of tracking technologies — GPS in smartphones, behavioral analytics in apps, biometric data in authentication systems — extends the invisible information gathering problem to nearly every interaction a person has with a digital financial service. What makes this ethically significant is not any single data point in isolation. It is the combination: location data layered over transaction data layered over behavioral data layered over third-party demographic profiles produces a portrait of a person's life that goes far beyond what anyone would recognize as having consented to when they downloaded a financial app.

How FinTech Firms Collect, Use, and Monetize Data

FinTech firms occupy a privileged position in the data economy. Because they process financial transactions, they collect data that is both highly sensitive and commercially valuable. Understanding how that data flows — how it is collected, used internally, and shared or sold externally — is essential for analyzing the ethical dimensions of FinTech business models.

How Data Is Collected

FinTech firms collect data through several channels, not all of which are visible to consumers. Direct collection occurs when consumers explicitly provide information — completing an account application, answering a risk questionnaire, entering payment details, or uploading identity documents. This data is collected with some degree of consumer awareness, though the full scope of how it will be used is rarely disclosed in terms most consumers understand.

Transactional collection occurs automatically as consumers use the platform. Every payment, trade, transfer, and inquiry generates a data record. Over time, these records reveal spending patterns, financial behaviors, life events, and economic circumstances in considerable detail. A consumer who uses a mobile payment app for daily purchases generates a transaction history that reveals where she shops, what she buys, how much she spends across different categories, when she is paid, and whether her financial situation is stable or deteriorating. This data is generated as a byproduct of the service rather than as a deliberate disclosure.

Behavioral collection captures how users interact with the platform itself — which features they access, how much time they spend on each screen, what notifications they respond to, and how their usage patterns change over time. This behavioral data is used for product improvement and personalization, but also for identifying users who may be experiencing financial stress, who may be susceptible to specific product offers, or whose engagement patterns suggest they are considering leaving the platform.

Third-party data acquisition supplements what firms collect directly. FinTech companies purchase data from data brokers, credit bureaus, and other sources to enrich their internal profiles. A consumer who provided only a name and email address to open an account may not know that the firm has purchased a comprehensive profile of her activities from a data broker — including inferred income, shopping behaviors, social media activity, and hundreds of other attributes — and incorporated it into the models that govern her experience on the platform.

Internal Data Use

Internally, FinTech firms use consumer data in ways that range from clearly beneficial to ethically contested. Credit decisioning is among the most consequential: algorithms that incorporate transaction data, behavioral signals, and third-party data to assess creditworthiness determine who gets access to credit and at what terms. When these algorithms work as intended, they expand credit access to people that traditional scoring models would have excluded. When they embed historical biases or use proxy variables that correlate with protected characteristics, they reproduce and amplify discrimination at scale.

Personalization is another significant internal use. FinTech platforms customize the products they offer, the notifications they send, and the pricing they display based on what they know about each user. Personalization can serve consumer interests — showing relevant products, providing timely alerts, adjusting communication frequency to match preferences. It can also serve the firm's interests at the consumer's expense — adjusting the prices or terms shown to individual consumers based on inferred willingness to pay, or presenting high-margin products prominently to consumers whose behavioral profiles suggest vulnerability to certain types of offers.

Fraud detection and risk management represent internal data uses that most consumers would accept as legitimate and beneficial. A payment platform that identifies transactions deviating from a customer's established patterns and flags them for review is using data in the customer's interest. The ethical question becomes more complex when the same behavioral models used for fraud detection are also applied to marketing, credit pricing, or customer retention without the consumer's awareness.

External Data Monetization

The most ethically contested dimension of FinTech data practice is external monetization — selling, sharing, or licensing consumer data to third parties. Data sales involve transferring consumer data to brokers, advertisers, or other businesses that use it for their own purposes. Data sharing involves providing consumer data to partners or affiliates as part of commercial arrangements. Targeted advertising involves using consumer data to deliver advertising within the platform, generating revenue from advertisers who pay for access to a precisely defined audience.

The ethical concern with external monetization centers on the gap between the context in which data was collected and the context in which it is used. A consumer who provided financial account information to receive a budgeting service did not consent, in any meaningful sense, to having her transaction data shared with an employer who uses it to screen job candidates, a health insurer who uses it to price her policy, or a political campaign that uses it to target her with advertising. The technical fact that consent may have been obtained through a terms-of-service agreement the consumer did not read does not make it ethically meaningful consent. Consent obtained through opacity is not the same as informed consent.

This distinction between formal consent and meaningful consent is one of the central tensions in data ethics. Formal consent — a checkbox, a terms-of-service agreement, a privacy policy — satisfies legal requirements and provides a defense against regulatory enforcement. Meaningful consent — disclosure that is clear, specific, and given when the consumer can reasonably understand what she is agreeing to — is a higher standard that formal consent mechanisms routinely fail to meet. Professionals who are serious about data ethics need to ask not just whether consent was technically obtained but whether it was genuinely informed.

The Census Bureau and the Consequences of Purpose Drift

Understanding why data ethics matters as a professional discipline — rather than a bureaucratic formality — is helped by examining historical cases where data collected for one purpose was used for another with serious consequences.

The U.S. Census Bureau collects detailed information about every person in the United States — age, race, income, family structure, and address. Respondents are legally required to participate and are told their responses are confidential and will be used only for statistical purposes. For most of American history, this confidentiality commitment was honored. But the Census Bureau has not always kept confidential the information it collected, and this became apparent after the attack on Pearl Harbor.

Following the Japanese attack on Pearl Harbor in December 1941, the federal government moved to identify, classify, and ultimately intern Japanese Americans living on the West Coast. The Census Bureau provided neighborhood-level data identifying concentrations of Japanese Americans — information that was then used to facilitate the forced removal and internment of over 120,000 people, the majority of them U.S. citizens. The census data, collected under a promise of confidentiality for statistical use, was repurposed for a surveillance and enforcement function that respondents could not have anticipated and would not have consented to.

The principle this case illustrates — purpose limitation — is foundational in data ethics. When people provide information in response to promises about how it will be used, they make a disclosure decision based on those promises. Using that information for different purposes violates the trust on which the original disclosure was based, regardless of whether the new purpose is viewed as legitimate by the party making the decision. The harm in the Japanese American internment case was catastrophic, but the underlying dynamic — data collected for one purpose, used for another — is the same dynamic that operates when a FinTech firm uses transaction data collected to process payments for credit pricing, health risk assessment, or targeted advertising without adequate disclosure and consent.

The ACM/IEEE Software Engineering Code of Ethics

The ethical responsibilities that attach to data in FinTech are not limited to corporate strategy and business model decisions. They also belong to the engineers, data scientists, and technologists who design the systems that collect and process personal data. The computing profession has developed a formal ethical framework to address these responsibilities.

The Software Engineering Code of Ethics and Professional Practice was developed jointly by the Association for Computing Machinery and the Institute of Electrical and Electronics Engineers — the two largest professional organizations in computing and engineering. The ACM/IEEE Code establishes eight principles governing how computing professionals should conduct their work. The paramount principle is that software engineers shall act consistently with the public interest — that obligation to the public takes priority over the interests of employers and clients when they conflict.

For FinTech professionals, the ACM/IEEE Code is relevant in several ways. It establishes that engineers bear professional responsibility for the systems they build and the consequences those systems produce. A data scientist who builds a credit scoring model that she knows uses proxy variables correlated with race has an obligation — not merely a legal one — to flag that problem and advocate for its correction. A software engineer who implements an invisible data collection feature without user notification is professionally responsible for that design choice, not simply the product manager who requested it.

The practical implication for FinTech is direct. Engineers and data scientists who build systems that collect personal data invisibly, that use behavioral signals consumers did not know were being evaluated, or that incorporate variables with discriminatory effects are not simply executing product requirements. They are making choices with ethical consequences, and the ACM/IEEE Code establishes that those choices carry professional responsibility. The code provides a framework for raising concerns, pushing back on harmful design decisions, and, when necessary, declining to build something that cannot be built ethically.

Putting the Frameworks to Work

The five ethical frameworks covered in this chapter are not ends in themselves — they are tools for analyzing real situations where competing interests, incomplete information, and genuine uncertainty make the right answer difficult to see. The Toysmart case introduced at the opening of this chapter illustrates how these tools work when pointed at a real problem.

A utilitarian analysis asks who is affected and how. The benefits of selling the database flow primarily to Toysmart's creditors, who recover more of the money they are owed. The harms fall on approximately 250,000 families whose personal data — collected under an explicit promise of non-disclosure — would be transferred to an unknown buyer for unspecified purposes.

Children's information is included. A complete utilitarian analysis requires honestly accounting for both sides: the number of people affected, the severity of the harm, and whether the aggregate outcome favors selling or not selling.

A deontological analysis asks whether the proposed sale treats affected parties as ends in themselves. Toysmart's customers provided their data in response to an explicit promise. Using that data in a way that directly contradicts the promise — regardless of the financial pressure driving the decision — fails the categorical imperative test. The customers are being used as a means to satisfy creditors' claims rather than treated as people whose reasonable expectations deserve respect.

A social contract analysis asks what the terms of the agreement between Toysmart and its customers were, and whether the proposed sale honors them. The TRUSTe seal and the explicit privacy policy created a clear commitment. Customers made a disclosure decision based on that commitment. Selling the data violates the implicit agreement regardless of what bankruptcy law might permit.

A virtue ethics analysis asks what an organization of genuine integrity would do. A company of good character, facing bankruptcy, would look for ways to honor its commitments even under financial pressure — seeking buyers willing to assume the privacy obligations, destroying the data rather than selling it, or finding other assets to satisfy creditors — rather than treating the privacy promise as an obligation that evaporates when it becomes inconvenient.

Stakeholder analysis maps who has interests at stake: creditors seeking recovery, customers whose promises should be honored, children whose information cannot be consensually transferred by the adults who trusted Toysmart on their behalf, privacy advocates representing broader public interests in data protection norms, and regulators charged with preventing deceptive practices.

These frameworks do not all point in the same direction with equal force, and working through where they converge and where they diverge is part of what makes the Toysmart case genuinely instructive. The ethical analysis this chapter equips is the first layer. Chapter 8 adds the second — the legal frameworks that complete the picture and allow the case to be analyzed in full.

When the Law Is Not Enough

The ethical frameworks covered in this chapter are powerful tools. But there is a limitation they share that FinTech professionals must understand before moving into the legal landscape that Chapter 8 covers: ethical obligation and legal compliance are not the same thing, and they do not always point in the same direction.

This gap between ethics and law runs in both directions. A company can act legally and still act unethically. It can also face ethical obligations in situations where the law provides no clear guidance at all.

The first direction is the more familiar one. Laws are written at specific moments in time, with specific harms in mind. They reflect the political compromises that made them passable and the practical limits of what legislators could anticipate. The result is that legal frameworks routinely lag behind technological and commercial practice. A FinTech firm that collects behavioral data from millions of users and sells it to insurers may be doing something entirely legal under current U.S. law. Whether it is ethical — whether it honors the reasonable expectations of the people whose data is being sold, whether it treats them as ends in themselves, whether it produces outcomes that a reasonable person would endorse — is a separate question that the law does not answer.

The second direction is less often acknowledged but equally important. Sometimes the law permits something that is ethically wrong. Sometimes a practice is technically legal but violates the purpose limitation, transparency, or consent principles that ethical frameworks require. And sometimes a firm faces choices where multiple legal obligations conflict — where honoring one legal duty appears to require violating another — and only ethical reasoning can provide a framework for navigating the tension.

For FinTech professionals, the ethics-law gap has a practical consequence: legal compliance cannot be the endpoint of ethical analysis. It must be the starting point. Asking "is this legal?" is a necessary question. It is not a sufficient one. The next question — "even if this is legal, is it right?" — requires the tools this chapter has covered, not a lawyer.

This distinction becomes concrete in the chapters that follow. Chapter 8 maps the specific legal frameworks that govern data privacy in the United States — the FCRA, GLBA, COPPA, the FTC's enforcement authority, and others. Those frameworks define what FinTech firms are required to do. What they are obligated to do as a matter of ethics may be more, less specific, or structured around considerations the law does not capture at all.

Understanding both layers — the legal floor and the ethical standard above it — and knowing how to reason about the space between them, is one of the core competencies this course is designed to develop.

Key Takeaways

Ethics is a rational examination of people's moral beliefs and actions. It is a branch of philosophy, more than 2,000 years old, concerned with determining what is good or bad and how people ought to behave. Ethics is distinct from legality — legal compliance is the floor, not the ceiling of ethical obligation. Professionals who treat anything legal as automatically acceptable are not engaging in ethical reasoning.

The major ethical frameworks relevant to data ethics are: utilitarianism, a consequentialist theory that evaluates actions based on whether they produce the greatest good for the greatest number; deontological ethics, which evaluates actions based on universal duties and the imperative to treat people as ends in themselves rather than means; virtue ethics, which focuses on the character and integrity reflected in an action rather than its rules or consequences; social contract theory, associated with Hobbes's observation that life without rules and enforcement is a state of nature, which grounds ethical obligations in reasonable expectations and mutual agreement; and stakeholder theory, which requires identifying all affected parties and considering the impact on each before making decisions.

Privacy is an ethical value that protects autonomy, dignity, safety, and trust. It is not simply a legal concept. Public records are those containing information reported to a government agency for the purpose of informing the public. The reasonable expectation of privacy — established in Katz v. United States in 1967 — holds that the government needs a court order to intrude upon communications in which a person has a reasonable expectation of privacy, and provides an ethical as well as legal standard for evaluating data practices. RFID technology, which uses wireless transmission to identify and track objects and individuals, and invisible information gathering — the automatic collection of data, such as IP addresses and browsing behavior, that individuals are unaware of — are two specific privacy-implicating technologies that FinTech professionals encounter regularly.

FinTech firms collect data through direct, transactional, behavioral, and third-party channels. Internal uses include credit decisioning, personalization, and fraud detection — each with varying degrees of consumer benefit and ethical risk. External monetization, which involves selling or sharing consumer data with third parties, raises the most significant ethical concerns about meaningful consent and the distance between what consumers were told and what is actually being done with their data.

The Census Bureau's use of census data to facilitate the internment of Japanese Americans after the attack on Pearl Harbor is the foundational historical illustration of purpose drift — using data collected for one purpose for a different, unanticipated, and unconsented purpose. The principle it establishes — that trust in data collection depends on honoring promises about use — is directly applicable to contemporary FinTech data practices.

The Software Engineering Code of Ethics and Professional Practice was developed by the Association for Computing Machinery and the Institute of Electrical and Electronics Engineers. It establishes that computing professionals have ethical obligations to the public that take priority over employer and client interests, and it provides an accountability framework for engineers and data scientists who build the systems that collect and process personal data. FinTech engineers who build invisible data collection systems, embed discriminatory variables, or implement privacy-hostile design choices bear professional responsibility for those decisions under the ACM/IEEE Code.

Legal compliance and ethical obligation are related but distinct. The law defines a floor — what is required and what is prohibited. Ethics asks a different question: what is right, regardless of what the law permits? This gap runs in both directions. A company can act legally and still act unethically, particularly when laws lag behind technological practice. And a company can face ethical obligations in situations where the law provides no guidance at all. For FinTech professionals, legal compliance is the starting point of ethical analysis, not the endpoint. The frameworks covered in this chapter — utilitarian, deontological, virtue-based, contractarian, and stakeholder — are the tools for answering the question that the law alone cannot resolve.

Chapter 8: U.S. Data Privacy Law and Consumer Protection

Introduction

Chapter 7 established the ethical framework for thinking about data privacy and introduced the gap between legal compliance and ethical obligation. This chapter turns to the legal side of that equation — what U.S. law actually requires of FinTech firms that collect, use, and share consumer data, and where the gaps in that legal framework leave consumers exposed and firms navigating uncertain terrain.

The United States does not have a single, comprehensive federal data privacy law. Unlike the European Union, which adopted the General Data Protection Regulation in 2018 as an omnibus framework governing the collection and use of personal data across all sectors, the United States has taken a sectoral approach: different laws apply to different industries, different types of data, and different categories of harm. Financial data is governed by certain rules. Children's data by others. Health data by still others. And large swaths of personal data — including much of what FinTech firms collect and monetize — fall under no specific federal privacy law at all, leaving the Federal Trade Commission's general consumer protection authority as the primary enforcement backstop.

This patchwork is not an accident. It reflects a deliberate policy choice — or, more precisely, a series of choices made across several decades — to regulate privacy reactively, addressing specific harms as they became politically visible rather than establishing proactive rights. The result is a legal landscape that is complex to navigate, uneven in its protections, and increasingly contested as the scale of data collection has grown far beyond what any of the existing frameworks anticipated.

This chapter maps that landscape. It covers the major federal privacy laws applicable to FinTech, the roles of the FTC and CFPB as the primary consumer protection enforcers, the specific provisions of the Fair Credit Reporting Act that FinTech firms must understand, the Carpenter v. United States decision and its implications for digital data, state privacy law with particular attention to California, and the EU's GDPR as the international benchmark against which U.S. law is measured. It closes by evaluating the gaps — what U.S. law does not require, why those gaps matter, and what FinTech professionals should understand about operating in a space where the law is still catching up to practice.

The Structure of U.S. Privacy Law

To understand U.S. data privacy law, you have to understand why it is structured the way it is. The sectoral approach — separate laws for separate sectors — reflects several characteristics of the American legislative tradition: a preference for addressing specific, demonstrated harms over anticipatory rights frameworks; the lobbying power of industries that resist broad privacy regulation; and a general constitutional framework that does not include an explicit right to privacy at the federal level.

The practical consequence of the sectoral approach is that the level of privacy protection a person receives depends almost entirely on the type of data involved and the industry collecting it. A patient's health records are protected by HIPAA. A student's educational records are protected by FERPA. A consumer's credit information is protected by the Fair Credit Reporting Act. But the same person's general financial transaction data — her payment history, her spending patterns, her income and saving behaviors — may be collected and used by a FinTech firm with minimal legal constraints beyond the general prohibition on deceptive practices enforced by the FTC.

This creates what privacy scholars call the contextual integrity problem: people share information in specific contexts, with reasonable expectations about how it will be used within that context. Health information shared with a doctor carries an expectation that it will be used for treatment. Financial information shared to process a payment carries an expectation that it will be used to complete the transaction. When information flows outside the context in which it was shared — when payment data is used for credit scoring, behavioral data is sold to insurers, or transaction history is used for employment screening — the person's contextual expectation is violated, even if no specific law prohibits the transfer.

The table below provides an overview of the major privacy laws and frameworks applicable to FinTech firms operating in the United States.

Law / Framework	Scope	Primary Enforcer	Key FinTech Relevance
Fair Credit Reporting Act (FCRA, 1970)	Consumer credit information; credit reports and credit reporting agencies	FTC; CFPB; private right of action	Governs credit bureau data used in FinTech lending and credit decisioning; imposes accuracy, access, and adverse action requirements
Gramm-Leach-Bliley Act (GLBA, 1999)	Financial institutions and their handling of nonpublic personal information	FTC; federal banking regulators; SEC; CFTC	Requires FinTech firms qualifying as financial institutions to provide privacy notices and safeguard customer data
Children's Online Privacy Protection Act (COPPA, 1998)	Online collection of personal information from children under 13	FTC	Applies to FinTech apps or services directed at or knowingly

			collecting data from children; requires verifiable parental consent
Electronic Communications Privacy Act (ECPA, 1986)	Electronic communications; stored communications; pen register data	DOJ; criminal prosecution	Governs government access to customer communications stored with FinTech platforms; relevant to data requests from law enforcement
California Consumer Privacy Act / CPRA (2018/2020)	Personal information of California residents; businesses above revenue/data thresholds	California Privacy Protection Agency; California AG	Most comprehensive U.S. state privacy law; de facto national standard for many firms; grants access, deletion, and opt-out rights
FTC Act Section 5 (1914, amended)	Unfair or deceptive acts or practices in commerce; applies to most businesses	FTC	Primary federal tool for enforcing privacy and data security commitments; used against companies that break privacy promises or fail to protect consumer data
EU General Data Protection Regulation (GDPR, 2018)	Personal data of EU residents; applies to any firm processing such data regardless of location	National Data Protection Authorities; European Data Protection Board	International benchmark; applies to U.S. FinTech firms serving EU customers; rights-based framework contrasts sharply

			with U.S. sectoral approach

The Federal Trade Commission

The Federal Trade Commission, or FTC, is the primary federal consumer protection agency in the United States. Its authority over data privacy and security derives primarily from Section 5 of the FTC Act, which prohibits unfair or deceptive acts or practices in or affecting commerce. The FTC is not a privacy regulator in the sense of having a comprehensive mandate to protect privacy as a right — it is a consumer protection agency with authority to act when companies' data practices harm consumers through deception or unfairness.

The Deception Standard

A deceptive act or practice under FTC law is one that involves a material representation, omission, or practice that is likely to mislead a consumer acting reasonably under the circumstances. In the privacy context, deception most commonly arises when a company makes promises about its data practices — in a privacy policy, a terms-of-service agreement, a marketing representation, or a third-party certification like the TRUSTe seal — and then fails to honor them.

The Toysmart case introduced in Chapter 7 was a deception case: Toysmart had made explicit promises that personal data would not be sold to third parties, and the proposed sale would have been directly contrary to those promises. The FTC's authority to intervene rested on its power to prevent deceptive practices, not on any specific data privacy statute requiring that the promises be kept.

This has an important implication: under the FTC's deception framework, the content of a privacy policy is not merely a disclosure — it is a legally binding commitment. A company that promises not to sell data and then sells it has deceived consumers. A company that promises to implement reasonable security measures and then fails to do so has deceived consumers. The practical lesson is that FinTech firms need to mean what their privacy policies say. Policies written to satisfy legal requirements but not actually followed are a source of legal and regulatory risk, not just ethical concern.

The Unfairness Standard

An unfair act or practice is one that causes or is likely to cause substantial injury to consumers that is not reasonably avoidable and not outweighed by countervailing benefits. The unfairness standard allows the FTC to act even when no specific promise was broken — when a data practice causes real consumer harm regardless of what the company disclosed.

The FTC has used its unfairness authority primarily in data security cases — actions against companies whose inadequate security practices led to data breaches that exposed consumer information to identity theft and fraud. A company that collects sensitive consumer financial data but fails to implement basic security measures is engaging in an unfair practice even if it never made any specific promises about security, because the failure to protect the data causes harm that consumers cannot reasonably avoid on their own.

Data security enforcement under the FTC's unfairness authority has produced a substantial body of consent orders — binding settlements in which companies agree to implement specific security measures and submit to independent audits for a period of years. These consent orders have effectively established a body of data security standards that, while not statutory law, carry significant practical weight as indicators of what the FTC considers reasonable practice.

The Limits of FTC Authority

The FTC's privacy enforcement authority has significant structural limitations that FinTech professionals need to understand. The FTC Act does not give the FTC the power to seek civil monetary penalties for first-time violations — it can issue cease-and-desist orders and require companies to implement remedial measures, but it cannot impose fines unless a company violates a prior consent order. This limitation significantly reduces the deterrent effect of FTC enforcement for companies that have not previously been sanctioned.

The FTC's rulemaking authority over privacy and data security has also been constrained by the lengthy and resource-intensive process required for formal rulemaking. While the FTC has issued rules in specific areas — including rules implementing the Children's Online Privacy Protection Act and rules governing the disposal of consumer financial information — it has not issued comprehensive privacy or data security rules applicable to FinTech broadly.

These limitations are relevant context for evaluating the adequacy of U.S. privacy law. When advocates argue that U.S. privacy regulation has significant gaps, one dimension of that argument is that the FTC, while a capable consumer protection agency, lacks the statutory tools to serve as a comprehensive privacy regulator in the way that European data protection authorities can under the GDPR.

The Consumer Financial Protection Bureau

The CFPB, covered in depth in Chapter 2, is the primary federal regulator of consumer financial products and services — the exact space where most FinTech companies operate. Its significance in this chapter is specific: unlike the FTC, whose privacy enforcement authority derives from general consumer protection law, the CFPB has direct supervisory authority over FinTech firms' data practices as part of its mandate to protect consumers in financial markets.

The CFPB has expanded its supervision to cover nonbank FinTech firms that pose risks to consumers, meaning it can examine a company's data practices without waiting for a specific complaint or harm to arise. Its enforcement actions in this space have focused on unauthorized access to consumer financial data, improper data sharing, and failures to honor consumers' data rights.

The most significant CFPB action affecting FinTech data practices is the Section 1033 rule, finalized in 2024 and introduced in Chapters 2 and 6. In the data privacy context, Section 1033 matters because it establishes affirmative consumer data rights — the right to access financial data in machine-readable form and to revoke third-party access at any time — and imposes data protection obligations on the third parties who receive that data. Those obligations include use limitations: authorized third parties cannot use consumer financial data for purposes beyond what the consumer authorized. Section 1033 represents the closest the United States has come at the federal level to the kind of data portability and access rights that characterize the GDPR, though it is narrower in scope.

The CFPB's enforcement posture has shifted significantly across administrations, and as of early 2026 the bureau was operating under constraints that had reduced its enforcement activity substantially. What does not shift with the CFPB's political fortunes are the underlying statutory authorities. FinTech firms that treat reduced enforcement activity as an invitation to reduce compliance rigor are taking a risk that extends beyond the current moment — the authority remains in place, and the FTC's concurrent enforcement jurisdiction over many of the same practices does not fluctuate with CFPB leadership changes.

The Fair Credit Reporting Act

The Fair Credit Reporting Act, enacted in 1970 and substantially amended since, is the primary federal law governing the collection, use, and sharing of consumer credit information. It is one of the most consequential consumer protection laws in the United States, because credit information affects consumers' access to loans, housing, employment, and a range of other life opportunities. For FinTech firms involved in lending, credit scoring, or the use of consumer financial data in underwriting, the FCRA is a foundational compliance obligation.

What the FCRA Requires

The Fair Credit Reporting Act does several important things. It gives consumers the right to request a free copy of their credit report every year from each of the three major credit bureaus — Equifax, Experian, and TransUnion. This annual free report right exists specifically so that consumers can review the accuracy of their credit information and identify errors or fraudulent accounts. It ensures that information about criminal convictions does not haunt a consumer's credit report forever — most negative information, including most criminal records, must be removed from credit reports after seven years, though certain convictions and bankruptcies have longer reporting periods. And it is designed to promote the accuracy of credit reports by giving consumers tools to dispute inaccurate information and requiring credit bureaus and furnishers to investigate disputes and correct errors.

These three provisions — the annual free report right, the criminal conviction time limit, and the accuracy promotion mandate — are the core consumer protections of the FCRA. They represent an important set of rights, and FinTech firms that use consumer credit data have obligations under each of them.

Permissible Purposes

The FCRA does not allow anyone to access a consumer's credit report for any purpose. It establishes a defined set of permissible purposes — circumstances under which a consumer reporting agency may furnish a consumer report. Permissible purposes include: a credit transaction initiated by the consumer; employment screening, subject to specific requirements including written consent; the underwriting of insurance; a legitimate business need in connection with a business transaction initiated by the consumer; and a few other specified uses.

The permissible purpose requirement is significant for FinTech because it means that accessing a consumer's credit report outside of these defined circumstances is a violation of federal law — regardless of what the FinTech firm believes its legitimate business purpose to be. FinTech firms that use credit data must have a clear permissible purpose for each access and must document it. Building a credit score as part of a lending application is a permissible purpose. Pulling a credit report to target advertising is not.

Adverse Action Notices

When a FinTech firm takes an adverse action against a consumer based in whole or in part on information in a consumer report — denying a loan application, offering credit on less favorable terms than requested, increasing an insurance premium — the FCRA requires it to provide the consumer with an adverse action notice. The notice must identify the consumer reporting agency that supplied the report, inform the consumer of their right to a free copy of the report within 60 days, and identify the principal factors that adversely affected the consumer's credit score.

The adverse action notice requirement is practically important for FinTech credit decisioning algorithms. If an algorithm denies a loan or worsens loan terms, the consumer is entitled to know the principal factors behind that decision and to obtain the credit report on which it was based. This creates a transparency obligation for algorithmic credit decisions that is often in tension with the opacity of sophisticated machine learning models — a tension that the CFPB and consumer advocates have highlighted as a specific regulatory concern.

The FCRA and FinTech Credit Decisioning

The FCRA was written before algorithmic credit decisioning existed in its current form. Its requirements were designed for a world in which credit bureaus compiled credit histories and lenders reviewed them to make manual underwriting decisions. The application of FCRA requirements to automated credit scoring models that incorporate hundreds of variables — including transaction data, behavioral signals, and alternative data sources that go well beyond the traditional credit file — raises interpretive questions that courts, regulators, and the industry are still working through.

The central question is whether the FCRA's definition of a consumer report — broadly, any communication by a consumer reporting agency bearing on a consumer's creditworthiness, credit standing, credit capacity, character, general reputation, personal characteristics, or mode of living — is broad enough to capture the alternative data products that FinTech lenders increasingly use. If a FinTech firm purchases a behavioral analytics score derived from a consumer's app usage, transaction patterns, and social media activity from a data vendor, and uses that score in a credit decision, is the vendor a consumer reporting agency subject to FCRA requirements? The answer is not obvious, and its resolution has significant implications for both consumer protection and FinTech business models.

The Gramm-Leach-Bliley Act

The Gramm-Leach-Bliley Act, or GLBA, enacted in 1999, established the financial privacy and data security framework that applies to financial institutions broadly — including, in many cases, FinTech firms. GLBA has three privacy-related components: the Privacy Rule, the Safeguards Rule, and the pretexting provisions.

The Privacy Rule

GLBA's Privacy Rule requires financial institutions to provide customers with clear notice of their information-sharing practices and to give consumers the opportunity to opt out of certain types of information sharing with nonaffiliated third parties. The privacy notice requirement means that customers must be told what nonpublic personal information the institution collects, with whom it shares the information, and how it protects the information. The opt-out right means that customers must be given the opportunity to prevent the institution from sharing their nonpublic personal information with nonaffiliated third parties for certain purposes.

The opt-out framework under GLBA is notably weaker than the opt-in frameworks that characterize stronger privacy regimes like the GDPR. An opt-out model requires consumers to take affirmative action to prevent data sharing — the default is sharing, not protection. Most consumers never exercise opt-out rights because they either do not receive the notice, do not read it, or do not understand the consequences of inaction. Privacy advocates have consistently argued that the opt-out model provides inadequate consumer protection precisely because it relies on consumer action that most consumers will not take.

The GLBA Privacy Rule applies to financial institutions — a term that includes not only banks and credit unions but also companies that are significantly engaged in financial activities, including many FinTech firms. A lending platform, a payment processor, and a robo-advisor may all qualify as financial institutions for GLBA purposes and must comply with its privacy notice and opt-out requirements.

The Safeguards Rule

GLBA's Safeguards Rule requires financial institutions to implement a written information security program that contains administrative, technical, and physical safeguards for customer information. The FTC and the federal banking regulators have both issued Safeguards Rule implementing regulations, with the FTC's version applying to nonbank financial institutions including many FinTech firms.

The Safeguards Rule was substantially updated in 2023 to align with current cybersecurity standards. The updated rule requires financial institutions to designate a qualified individual to oversee the information security program, conduct a risk assessment, implement specific safeguards including encryption, access controls, and multi-factor authentication, and notify the FTC within 30 days of discovering a security breach affecting 500 or more customers. For FinTech firms that previously had light-touch security programs, the updated Safeguards Rule represents a meaningful compliance obligation with specific, enforceable requirements.

Children's Online Privacy Protection Act

The Children's Online Privacy Protection Act, or COPPA, enacted in 1998, governs the online collection of personal information from children under the age of 13. It requires operators of websites and online services directed to children, or that have actual knowledge they are collecting personal information from children, to provide clear notice of their data collection practices, obtain verifiable parental consent before collecting personal information from children, and provide parents with the ability to review, correct, and delete their child's information.

COPPA is enforced by the FTC and carries civil penalties of up to approximately $51,000 per violation. The FTC has brought dozens of enforcement actions under COPPA, including actions against large technology companies and online services. Google agreed to pay $170 million in 2019 to settle COPPA allegations related to YouTube's collection of data from children without parental consent — the largest COPPA penalty to date at the time.

For FinTech, COPPA is directly relevant in several contexts. A FinTech platform that allows minors to use its services — whether a payment app, a savings product, or an investment platform designed for young people — must comply with COPPA if it collects personal information from children under 13. The Toysmart case involved a company that had collected personal information about children as part of its family profiles, making COPPA directly applicable to the question of what could be done with that data.

COPPA's verifiable parental consent requirement is particularly demanding in digital contexts. It requires that consent be obtained through a mechanism that reasonably ensures the consenting party is the parent — a standard that simple checkbox consent does not meet. The FTC has recognized credit card verification, signed consent forms, and video calls as acceptable verification mechanisms, but the compliance burden is substantial and has led many general-audience platforms to take a categorical approach: prohibiting users under 13 entirely rather than building the parental consent infrastructure COPPA requires.

The Fourth Amendment and Digital Data: Carpenter v. United States

The constitutional dimension of data privacy in the United States derives primarily from the Fourth Amendment, which protects against unreasonable searches and seizures. For most of American history, Fourth Amendment doctrine tracked physical spaces — the government could not search your home without a warrant. As electronic communications, digital records, and mobile devices became the primary repositories of personal information, courts were forced to determine how constitutional privacy protections apply to data.

The U.S. Supreme Court's decision in Carpenter v. United States, decided in 2018, represents the most significant Fourth Amendment ruling in the digital privacy context. The case involved law enforcement officers who obtained months of cell-site location information — records held by Timothy Carpenter's wireless carrier showing the location of his phone at various times — without obtaining a warrant. The government argued that no warrant was required because the location records were voluntarily shared with a third party — the phone carrier — and therefore not protected by the Fourth Amendment under the so-called third-party doctrine.

The Supreme Court rejected that argument in a 5-4 decision written by Chief Justice Roberts. The Court held that accessing cell phone location data without a search warrant is a violation of the Fourth Amendment. The Court recognized that cell-site location information is qualitatively different from the types of third-party records that earlier cases had held unprotected — it provides a comprehensive, detailed record of a person's movements over time, revealing the privacies of life in a way that earlier technologies did not. The third-party doctrine, which had been developed in cases involving bank records and phone numbers, was not adequate to handle the pervasive surveillance capabilities created by modern digital technology.

The Carpenter decision is significant for FinTech for several reasons. First, it establishes that the comprehensiveness and granularity of digital records matter to constitutional analysis — the more complete a surveillance capability a dataset provides, the stronger the Fourth Amendment protections that attach to it. Second, it signals that courts are willing to adapt constitutional doctrine to new technological realities rather than mechanically applying rules developed for different circumstances. Third, it has implications for law enforcement requests for FinTech data: a warrant may be required to access digital financial records if those records provide the kind of comprehensive personal history that Carpenter addressed.

It is important to note that Carpenter addresses government access to data — it governs what law enforcement can do, not what private companies can do. FinTech firms are not constrained by the Fourth Amendment in their own data collection practices. But Carpenter matters as a signal about how courts think about digital privacy and as a legal boundary on the government's ability to access the data FinTech firms collect from their customers.

State Privacy Law

In the absence of a comprehensive federal privacy law, states have increasingly filled the gap. The result is a growing patchwork of state privacy laws that applies to FinTech firms serving customers across the country — creating compliance complexity that rivals the regulatory fragmentation challenges discussed in Chapter 2.

The California Framework

California has been the most influential state actor in U.S. privacy law, and understanding its framework is essential for any FinTech firm operating at national scale. The California Consumer Privacy Act, or CCPA, enacted in 2018 and substantially expanded by the California Privacy Rights Act, or CPRA, in 2020, provides California residents with rights that significantly exceed what federal law requires.

Under the CCPA/CPRA framework, California residents have the right to know what personal information a business has collected about them and how it is being used; the right to request deletion of their personal information, subject to certain exceptions; the right to opt out of the sale of their personal information to third parties; the right to opt out of the sharing of their personal information for cross-context behavioral advertising; the right to correct inaccurate personal information; and the right to limit the use of sensitive personal information, including financial account information, for secondary purposes.

The CPRA also created the California Privacy Protection Agency — a dedicated state privacy enforcement body — and established the concept of sensitive personal information as a category warranting heightened protection. Financial account information, including account numbers, debit and credit card numbers, and access credentials, is classified as sensitive personal information, meaning that its secondary use for purposes beyond the primary transaction requires specific consumer authorization.

For FinTech firms, the California framework is effectively a national compliance standard: any firm with significant consumer operations in the United States is likely serving California residents and must comply with CCPA/CPRA as a practical matter. The cost of maintaining California-specific privacy practices that differ from national practices has led many large firms to extend California's consumer rights to all U.S. customers rather than managing state-by-state variation.

The Growing State Patchwork

Following California's lead, a growing number of states have enacted their own comprehensive privacy laws — including Virginia, Colorado, Connecticut, Texas, and others. While these laws share some common features, they differ on key dimensions: which businesses are covered, what consumer rights are provided, how opt-in versus opt-out consent works, and what enforcement mechanisms apply. The result is a compliance landscape in which FinTech firms must navigate overlapping and sometimes inconsistent state requirements in the absence of a federal standard that would preempt them.

The state patchwork argument is one of the primary industry arguments for federal privacy legislation: a single national standard would reduce compliance complexity and provide clearer consumer protections than a fragmented state-by-state framework. Consumer advocates respond that any federal law must be strong enough to provide genuine protections — a weak federal standard that preempts stronger state laws would leave consumers worse off, not better. This debate over federal preemption is one of the central fault lines in current privacy policy discussions.

The EU General Data Protection Regulation

The European Union's General Data Protection Regulation, known as the GDPR, came into effect in May 2018 and has become the most influential privacy law in the world. Its influence extends well beyond Europe: any company that collects or processes personal data of EU residents — regardless of where the company is located — must comply with the GDPR. For U.S. FinTech firms that have European customers, European partners, or European operations, GDPR compliance is a real operational requirement. And for U.S. privacy policy discussions, the GDPR serves as the most prominent benchmark against which the adequacy of U.S. law is measured.

The GDPR's Core Framework

The GDPR takes a fundamentally different approach to privacy than U.S. law. Where U.S. law is sectoral — addressing specific industries and specific types of harm — the GDPR is omnibus, applying to the collection and use of any personal data in any context. Where U.S. law is reactive — requiring action primarily after harm occurs — the GDPR is proactive, establishing affirmative rights and obligations that apply before any harm arises. And where U.S. law generally treats data protection as a consumer protection concern — a matter of preventing deception and unfairness — the GDPR treats privacy as a fundamental right, grounding its framework in the EU Charter of Fundamental Rights.

The GDPR's core requirements center on six principles for the lawful processing of personal data. Personal data must be processed lawfully, fairly, and transparently — meaning that individuals must know what data is being collected and on what legal basis. Data must be collected for specified, explicit, and legitimate purposes and not further processed in a way incompatible with those purposes — the purpose limitation principle that the Census Bureau case in Chapter 7 illustrated so vividly. Data collection must be limited to what is necessary for the stated purpose — data minimization. Data must be accurate and kept up to date. Data must not be kept longer than necessary for the purpose — storage limitation. And data must be processed with appropriate security.

GDPR Rights and Their U.S. Contrast

The GDPR grants individuals a set of enforceable rights over their personal data that are substantially stronger than what U.S. law provides in most contexts. The right of access allows individuals to obtain a copy of all personal data held about them and information about how it is being used. The right to erasure — sometimes called the right to be forgotten — allows individuals to request deletion of their personal data in certain circumstances, including when the data is no longer necessary for the purpose it was collected or when consent is withdrawn. The right to data portability allows individuals to receive their personal data in a structured, machine-readable format and to transfer it to another service provider. And the right to object allows individuals to stop certain types of data processing, including processing for direct marketing purposes, with no questions asked.

Comparing these rights to what U.S. law provides illustrates the gap clearly. Under U.S. federal law, there is no general right to know what personal data a company holds about you. There is no general right to request deletion. There is no general right to data portability. The FCRA provides specific rights over credit information. COPPA provides specific rights for parents over children's data. The CFPB's Section 1033 rule provides specific rights over financial account data. But these are islands of protection in a sea of unregulated data collection. The GDPR provides a floor of protection that applies across all personal data, in all contexts, regardless of the industry or the type of information.

Lawful Basis and Consent

One of the most practically significant differences between U.S. and GDPR frameworks is the lawful basis requirement. Under the GDPR, every instance of personal data processing must rest on one of six lawful bases: the data subject's consent; the performance of a contract; compliance with a legal obligation; protection of vital interests; performance of a task in the public interest; or the legitimate interests of the controller, balanced against the individual's rights. Consent, where relied upon, must be freely given, specific, informed, and unambiguous — a pre-ticked box does not constitute valid consent under the GDPR.

This requirement eliminates the gap between formal consent and meaningful consent that Chapter 7 identified as a central ethical concern. Under the GDPR, a terms-of-service agreement that most consumers will not read is not a sufficient basis for processing their data for secondary commercial purposes. A company that wants to sell consumer data to third parties must either have explicit, informed consent for that specific purpose or demonstrate that it falls under one of the other lawful bases — which data sales typically do not.

The contrast with U.S. practice is stark. U.S. FinTech firms routinely rely on lengthy privacy policies and terms-of-service agreements to establish consent for data practices that most consumers are unaware of. Under GDPR standards, those same consent mechanisms would be legally inadequate for most secondary data uses. This gap is one reason why data-intensive FinTech companies have been cautious about expanding European operations — the compliance cost of genuinely meeting GDPR consent standards for the same practices that are routine in the U.S. market is substantial.

Enforcement and Penalties

The GDPR's enforcement mechanism is significantly stronger than the FTC's. Data protection authorities in each EU member state have the power to impose fines of up to 20 million euros or four percent of a company's global annual revenue, whichever is higher, for serious violations. These are not theoretical numbers: Meta has been fined over one billion euros for GDPR violations related to data transfers between Europe and the United States. Amazon was fined 746 million euros by Luxembourg's data protection authority. The scale of potential penalties creates genuine compliance incentives that the FTC's consent order regime, with its limited monetary sanctions for first-time violations, does not replicate.

The Gaps in U.S. Privacy Law and Their Consequences

Having mapped the legal landscape, it is worth stepping back to evaluate its adequacy — to ask where the gaps are, what consequences they have for consumers and for FinTech firms, and what the professional implications are for people working in this field.

What U.S. Law Does Not Require

U.S. federal law does not require FinTech firms to tell consumers what personal data they hold about them. It does not give consumers a general right to request deletion of their data. It does not require that data collection be limited to what is necessary for the stated purpose. It does not require meaningful consent — as opposed to formal disclosure — for secondary data uses. It does not provide a general right to opt out of data sales. And it does not impose the kind of comprehensive security requirements that would match the sensitivity and scale of the financial data FinTech firms collect.

These gaps are not abstractions. They have concrete consequences for the consumers whose data flows through FinTech platforms. A consumer whose transaction data is sold to a health insurer that uses it to price her policy has no federal right to know that the sale occurred, no right to demand deletion of the data, and no right to prevent the use. A consumer whose behavioral data is incorporated into a credit scoring model that denies her a loan has limited visibility into what data was used and limited ability to challenge its accuracy. A consumer whose financial data is held by a FinTech firm that fails to implement adequate security has no guarantee of meaningful compensation if a breach occurs.

The Compliance Implications for FinTech Firms

For FinTech professionals, the gaps in U.S. privacy law create both operational challenges and strategic risks. The operational challenge is navigating a complex, multi-layered compliance environment — FCRA,

GLBA, COPPA, FTC Act, state laws, and, for firms with international operations, GDPR — without a unified framework that resolves conflicts and fills gaps. The resources required to build and maintain compliance programs across this landscape are substantial, and the consequences of getting it wrong can include enforcement actions, civil litigation, and reputational damage.

The strategic risk is that the legal landscape will change. The patchwork of state laws will continue to grow. Federal privacy legislation, while it has failed to pass for years, remains a perennial legislative priority. The CFPB's Section 1033 framework continues to develop. Courts will continue to extend Carpenter's logic to new categories of digital data. FinTech firms that have built business models around data practices that are currently legal but ethically contested are exposed to the risk that the law will eventually catch up with the ethics — and that the transition will be disruptive.

This is precisely why Chapter 7's ethical framework matters alongside this chapter's legal map. Professionals who understand only the current legal requirements are calibrating to a moving target. Professionals who understand the ethical principles underlying privacy protection — consent, purpose limitation, data minimization, transparency — are better positioned to anticipate where the law is heading and to build practices that will remain defensible as the regulatory environment evolves.

The ToySmart Case Study:

Privacy Promises, Bankruptcy, and the Sale of Customer Data

The Case:

In 1999, Toysmart.com launched as an online toy retailer. To personalize the shopping experience, the company collected detailed personal information from its customers — names, addresses, family profiles, and the ages, interests, and wish lists of their children. Toysmart's privacy policy stated explicitly that this information would never be sold or shared with third parties. The company also displayed the TRUSTe privacy seal, a recognized third-party certification of responsible data practices.

In May 2000, Toysmart filed for bankruptcy. Its physical assets had limited value. Its customer database — profiles of approximately 250,000 families, including children's personal information — was the most valuable asset in the estate. Toysmart proposed to sell it.

The Federal Trade Commission filed a complaint to block the sale, arguing that selling data collected under an explicit promise of non-disclosure would constitute an unfair and deceptive trade practice under the FTC Act. Consumer advocates added that the database contained children's personal information subject to COPPA's strict transfer restrictions. Toysmart's creditors — vendors, investors, and others owed money — argued that the database was a legitimate estate asset and that the bankruptcy trustee had a fiduciary duty to maximize their recovery.

The case never reached a full ruling on the merits. The FTC and Toysmart reached a settlement permitting sale only to a buyer in a substantially similar line of business that agreed to assume Toysmart's privacy obligations. No such buyer emerged, and the database was eventually destroyed.

<u>The Central Question:</u>

Should the bankruptcy court have permitted Toysmart to sell its customer database to satisfy creditor claims, given the company's explicit privacy promises and the nature of the data involved?

Legal Frameworks at Play:

The following laws are directly relevant to this case. You covered all of them in Chapters 7 and 8.

The FTC Act, Section 5 prohibits unfair or deceptive acts or practices in commerce. Under the FTC's deception standard, a privacy policy is a binding public commitment. Breaking it — by selling data a company promised never to sell — constitutes a deceptive practice. The FTC's unfairness standard may also apply if the sale causes substantial consumer harm that consumers could not reasonably avoid.

COPPA governs the collection and transfer of personal information from children under 13. It requires verifiable parental consent and imposes strict limits on disclosing children's data to third parties. The Toysmart database included children's personal information collected under specific parental consent to Toysmart's stated practices — not to an unknown buyer.

Federal bankruptcy law obligates a bankruptcy trustee to maximize recovery for creditors from the assets of the estate. This duty is not optional. However, bankruptcy does not permit a sale that violates applicable law — including consumer protection law.

COPPA and the FTC Act do not automatically resolve the conflict with bankruptcy law. Both sides have legitimate legal grounding. That tension is the center of this case.

Your Tasks:

This case has three tasks. Complete them in order. You are being evaluated on the quality of your legal and ethical reasoning — not on which side you rule for in Task 3.

Task 1 — Argue for the Creditors

You represent Toysmart's creditors. Write a structured argument for why the bankruptcy court should permit the sale of the database. Your argument must address both the legal and ethical case for the sale. Engage with the strongest counterarguments and explain why they do not prevail.

Task 2 — Argue for the Customers

You represent Toysmart's customers, in coordination with the FTC. Write a structured argument for why the court should prohibit the sale. Your argument must address both the legal and ethical case against the sale — including the specific issues raised by children's data under COPPA. Engage with the creditors' strongest arguments and explain why they do not prevail here.

Task 3 — Render Your Judgment

You are the judge. Rule on whether the sale should be permitted, prohibited, or permitted only under specific conditions you define. Your ruling must: identify which legal framework controls the outcome and why; address the relationship between the legal analysis and the ethical analysis, including any gap between what the law permits and what ethics requires; and defend your ruling against the strongest objection to it.

Key Takeaways

The Federal Trade Commission enforces privacy and data security obligations primarily through Section 5 of the FTC Act, which prohibits unfair or deceptive acts or practices. Under the deception standard, a privacy policy is a legally binding commitment — breaking it is deceptive. Under the unfairness standard, inadequate data security that causes consumer harm is actionable even without a broken promise. The FTC's authority is limited: it cannot impose civil monetary penalties for first violations, and its rulemaking authority is constrained.

The Consumer Financial Protection Bureau has authority over consumer financial products and services, including the data practices of FinTech firms. The CFPB's Section 1033 rule, finalized in 2024, establishes consumer rights to access their financial data and to authorize third parties to access it — representing a meaningful step toward the data portability rights that characterize stronger privacy regimes. The CFPB's enforcement posture has varied significantly across administrations; the underlying statutory authority remains in place regardless of current enforcement priorities.

The Fair Credit Reporting Act governs consumer credit information. Its key provisions include: giving consumers the right to a free copy of their credit report every year; ensuring that criminal convictions do not permanently haunt credit reports; promoting the accuracy of credit reports through dispute rights and furnisher obligations; limiting access to credit reports to defined permissible purposes; and requiring adverse action notices when credit decisions are made based on consumer report information. These provisions apply to FinTech firms that use consumer credit data in their products and underwriting.

The Gramm-Leach-Bliley Act requires financial institutions — including many FinTech firms — to provide privacy notices, give consumers opt-out rights over certain data sharing, and implement written information security programs under the Safeguards Rule. The updated Safeguards Rule, effective in 2023, requires specific security measures including encryption, multi-factor authentication, and breach notification.

The Children's Online Privacy Protection Act requires verifiable parental consent before collecting personal information from children under 13 and applies to FinTech platforms directed at or knowingly serving children. It was directly implicated in the Toysmart bankruptcy because Toysmart's database included personal information about children collected under explicit privacy promises that the proposed sale would have broken.

In Carpenter v. United States, the Supreme Court held that accessing cell phone location information without a search warrant is a violation of the Fourth Amendment. The decision establishes that the comprehensiveness of a digital dataset matters to its constitutional protection and signals courts' willingness to adapt constitutional doctrine to new digital surveillance capabilities.

California's Consumer Privacy Act and California Privacy Rights Act provide California residents with rights to access, delete, correct, and limit the use of their personal information — rights that significantly exceed current federal law requirements and that many national FinTech firms extend to all U.S. customers as a practical compliance matter. A growing number of states have enacted comparable frameworks, creating a patchwork that complicates compliance in the absence of a federal standard.

The EU's General Data Protection Regulation is the most comprehensive and influential privacy law in the world. It applies to U.S. FinTech firms that process EU residents' personal data and requires: a lawful basis for every data processing activity; purpose limitation; data minimization; individual rights including access, erasure, portability, and the right to object; meaningful consent rather than formal disclosure; and enforcement through fines up to four percent of global annual revenue. The GDPR represents the primary international benchmark against which U.S. privacy law is measured — and the comparison reveals significant gaps in U.S. consumer protection.

The gaps in U.S. privacy law create both compliance challenges — navigating a complex multi-law environment without a unified framework — and strategic risks, as state laws continue to proliferate and the regulatory environment continues to evolve. Professionals who understand the ethical principles underlying privacy protection are better positioned to anticipate regulatory change than those who calibrate only to current legal requirements.

Chapter 9: Cybersecurity in FinTech — Threats, Vulnerabilities, and Defenses

Introduction

Picture a customer at a coffee shop using her laptop to pay her monthly bills over the complimentary Wi-Fi. In the background, an attacker on the same network is intercepting her unencrypted traffic, capturing her login credentials, and systematically accessing her financial accounts. She has no idea it is happening. The attacker does not need sophisticated tools or specialized knowledge — freely available software does most of the work. By the time she gets home, her accounts have been accessed, transactions have been initiated, and her financial data has been harvested for future use or sale on criminal marketplaces.

This scenario is not hypothetical. It represents the everyday operational reality of cybersecurity in financial services — a landscape in which the attack surface is vast, the adversaries are skilled and motivated, the data at stake is among the most sensitive that exists, and the consequences of failure are immediate and severe. Financial institutions have always been targets for theft. What has changed is the scale, the sophistication, and the sheer accessibility of the tools available to attackers.

FinTech firms occupy a particularly exposed position in this landscape. They are digital-first by design — their products and services are delivered through the same channels that attackers exploit. They often scale faster than their security programs can keep pace with. They rely on complex networks of third-party APIs, cloud infrastructure, and partner integrations that each introduce potential vulnerabilities. And they collect and process financial data of exactly the kind that attackers most want to steal, sell, or ransom.

This chapter builds the technical literacy that cybersecurity analysis in FinTech requires. It begins with the history of hacking — because understanding how the threat landscape evolved helps explain why it looks the way it does today. It then maps the major categories of threats facing FinTech firms, explains the attack tools and techniques that adversaries use, covers the defenses that firms deploy to protect themselves, examines the specific vulnerabilities created by FinTech's operating model, and describes what a firm should do when an attack occurs. Chapter 10 takes up the legal and policy dimensions — what the law requires, how information sharing works, and the debate over the Cybersecurity Information Sharing Act. This chapter provides the technical foundation that Chapter 10's legal analysis requires.

The History of Hacking

The term hacker did not begin as a description of a criminal. In the early days of computing — from the late 1950s through the early 1970s — a hacker was a creative programmer who wrote very elegant or clever programs. The term originated at MIT and other early computing centers to describe programmers who found innovative solutions to technical problems, who pushed machines beyond their intended capabilities, and who took genuine intellectual pleasure in the craft of programming itself. Hacking, in this original sense, was a term of respect — it described a particular kind of technical virtuosity.

The meaning of the term began to shift in the late 1970s and early 1980s as personal computers became available and networks began to connect them. A subculture emerged around phone phreaking — the practice of exploiting vulnerabilities in telephone networks to make free long-distance calls and explore network infrastructure. Phone phreaks like John Draper, who discovered that a toy whistle included in Cap'n Crunch cereal produced the precise 2600-hertz tone needed to access AT&T long-distance trunks, were not motivated primarily by financial gain. They were motivated by curiosity, by the challenge, and by the culture of technical exploration that the early computing community had established.

The 1980s brought the first widely publicized unauthorized computer intrusions — cases in which individuals accessed systems they had no permission to enter, reading files, leaving messages, and demonstrating vulnerabilities. The movie WarGames in 1983, depicting a teenage hacker who nearly triggers nuclear war by accessing a military computer, captured public anxiety about the new threat even as it misrepresented the actual capabilities and motivations of most hackers at the time. Real incidents — including the 1988 Morris Worm, one of the first pieces of self-replicating malicious code to spread across the internet, causing significant disruption to connected systems — began to demonstrate that the potential for serious harm was real.

The 1990s brought the commercialization of the internet, which transformed the hacking landscape fundamentally. Where early hacking had been largely the province of technically sophisticated individuals motivated by curiosity or challenge, the internet created both a massive new attack surface and a financial incentive structure that attracted organized criminal groups. Credit card theft, identity fraud, and the theft of proprietary business information became profitable enterprises. Hacking tools became commoditized — packaged into kits that required far less technical skill to deploy than the custom exploits of earlier eras.

By the 2000s, the threat landscape had evolved into three overlapping domains. Financially motivated criminal organizations — many operating from Eastern Europe and Russia — were conducting large-scale fraud operations targeting financial institutions and their customers. Nation-state actors — governments using cyberattacks as instruments of intelligence collection, economic espionage, and geopolitical competition — were targeting critical infrastructure and high-value corporate targets. And hacktivists — individuals and groups using hacking to promote political causes — were attacking organizations they perceived as harmful, exposing sensitive information, and disrupting operations to make political points.

Understanding this history matters for FinTech professionals because the threat environment today is a product of all three of these streams. The financial sector faces all categories of attacker simultaneously — organized criminals pursuing straightforward theft, nation-state actors pursuing intelligence or infrastructure disruption, and hacktivists targeting firms associated with controversial policies or practices. Each category of attacker uses different tools, has different objectives, and requires different defensive responses.

The Threat Landscape: Categories of Attack

Cybersecurity threats can be categorized in multiple ways — by the attacker's identity, by the attack method, by the target, or by the objective. For practical FinTech security analysis, the most useful framework combines the attack method with the attacker's general profile, because the appropriate defensive response depends on both. The table below provides an overview of the major threat categories facing FinTech firms.

Threat Category	Description	Primary FinTech Targets	Example
Phishing / Spear Phishing	Deceptive communications that trick recipients into revealing credentials or clicking malicious links; spear phishing is targeted at specific individuals or roles	Employees with system access; customers; executives	Email appearing to be from a payment processor requesting credential verification
Ransomware	Malware that encrypts a victim's data and demands	Core banking systems; customer data repositories;	2021 CNA Financial ransomware

	payment for the decryption key	payment infrastructure	attack; $40M ransom paid
Social Engineering	Psychological manipulation of people into performing actions or disclosing information; exploits human trust rather than technical vulnerabilities	Help desk staff; customer service; employees with privileged access	Attacker posing as IT support to reset credentials
Insider Threats	Malicious or negligent actions by current or former employees, contractors, or partners with authorized system access	Customer data; transaction systems; proprietary algorithms	Employee exfiltrating customer data before departure; negligent misconfiguration
Distributed Denial of Service (DDoS)	Flooding a target system with traffic to make it unavailable to legitimate users	Payment processing infrastructure; trading platforms; customer-facing apps	Attacks on major payment processors disrupting transaction processing
API Attacks	Exploitation of vulnerabilities in application programming interfaces that connect FinTech services to each other and to partner institutions	Open banking data connections; payment APIs; third-party integrations	Unauthorized access to customer account data through poorly secured API endpoints
Supply Chain / Third-Party Attacks	Compromising a vendor or partner to gain access to the primary target's	Any FinTech firm relying on third-party software or services	2020 SolarWinds attack; attackers compromised software update mechanism to

	systems through a trusted connection		reach thousands of organizations
Nation-State Attacks	Sophisticated attacks sponsored or directed by foreign governments targeting financial infrastructure for intelligence, disruption, or theft	Central financial infrastructure; large banks; payment networks	North Korean Lazarus Group attacks on SWIFT messaging system
Hacktivism	Use of hacking to promote a political cause; attackers may seek to disrupt operations, expose information, or deface public-facing systems	Firms associated with controversial policies or perceived social harm	Anonymous attacks on financial firms following political controversies

Phishing and Social Engineering

Phishing is among the most prevalent and consistently effective attack methods in the cybersecurity threat landscape. A phishing attack uses deceptive communications — most commonly email, but also SMS text messages (smishing) and voice calls (vishing) — to trick recipients into revealing credentials, clicking malicious links, or downloading malware. The communications are designed to appear legitimate, mimicking the visual identity and language of trusted institutions: banks, payment processors, internal IT departments, or regulatory agencies.

Spear phishing is a targeted variant in which the attacker customizes the deceptive communication for a specific individual or organization, using information gathered from social media, professional profiles, and public records to make the message more convincing. A spear phishing email addressed to a FinTech firm's accounts payable manager, appearing to come from the firm's CEO and requesting an urgent wire transfer, is an example of business email compromise — one of the most financially damaging forms of phishing attack. The FBI's Internet Crime Complaint Center has consistently identified business email compromise as one of the costliest forms of cybercrime, with aggregate losses in the billions of dollars annually.

Social engineering is the broader category of which phishing is a subset — it encompasses any technique that exploits human psychology rather than technical vulnerabilities to gain unauthorized access or information. Social engineering attacks exploit trust, authority, urgency, and fear. An attacker who calls a help desk claiming to be a locked-out employee and uses social pressure to get credentials reset is engaging in social engineering. An attacker who gains physical access to a secure facility by following an authorized employee through a secure door — a technique called tailgating — is engaging in social engineering. The common thread is that the attacker is manipulating people rather than exploiting code.

Social engineering is particularly difficult to defend against because it targets the human element of security — and humans are inherently susceptible to the psychological mechanisms these attacks exploit. An employee who receives an email that appears to come from her CEO, requests urgent action, and includes enough specific detail about the firm to seem plausible may comply even if she has been trained in phishing awareness. Security professionals sometimes describe this as the weakest link problem: a security system is only as strong as its least secure component, and in many organizations the least secure component is the people.

Ransomware

Ransomware is malicious software that encrypts a victim's data and displays a demand for payment — typically in cryptocurrency — in exchange for the decryption key. When a ransomware attack is successful, the victim organization loses access to its data entirely: files are encrypted with keys only the attacker holds, and systems become inoperable. The financial impact includes not just any ransom paid but the cost of operational disruption, recovery efforts, regulatory penalties, and reputational damage.

Ransomware attacks against financial services firms have been among the most costly and disruptive cyberattacks of the past decade. CNA Financial, one of the largest commercial insurers in the United States, paid a $40 million ransom in 2021 following an attack that disrupted its operations for weeks. The attack on the Colonial Pipeline in the same year, while not a financial firm, demonstrated the operational disruption that ransomware can cause to critical infrastructure — a category that includes payment networks and financial market infrastructure.

For FinTech firms, ransomware is a particularly acute threat because their operations depend entirely on the availability of their systems and data. A payment processing platform that cannot access its transaction data cannot process payments. A lending platform that cannot access its credit decisioning system cannot originate loans. A trading platform that cannot access its order management system cannot execute trades. The operational dependency on data availability means that even a temporary ransomware disruption can be catastrophically costly — which is precisely what makes it an effective extortion tool.

Insider Threats

Internal hackers — employees, contractors, and former employees with authorized access to systems — represent one of the most significant and underappreciated threats to FinTech security. They are a major problem for a specific reason that distinguishes them from external attackers: they are already authorized and have privileges within the system. An external attacker must first penetrate the organization's perimeter defenses, avoid detection, and escalate privileges. An insider already has access — their malicious activity can be harder to detect precisely because it looks, at least initially, like legitimate use.

Insider threats take two forms. Malicious insiders are employees or contractors who deliberately misuse their access — stealing customer data to sell, exfiltrating proprietary information before departing for a competitor, manipulating financial records, or facilitating external fraud by providing account credentials or bypassing controls. Negligent insiders are employees who do not intend harm but whose careless actions create vulnerabilities — clicking phishing links, using weak passwords, misconfiguring systems, or improperly handling sensitive data.

A characteristic that makes insider threats particularly concerning is that when criminals gain unauthorized access through external means, they typically make copies of data rather than removing the original — meaning the data owners can still use the original data while the attacker also possesses it. This pattern of silent, invisible exfiltration makes insider data theft difficult to detect through conventional monitoring that looks for system disruption or data destruction. The data is still there; it has simply been copied and taken.

Financial services firms have historically experienced insider threats across multiple dimensions. Traders at banks have manipulated records to conceal losses. Customer service representatives have accessed account data and sold it to fraud rings. Former employees have retained access credentials after departure and used them to access systems. The scale of customer data that FinTech platforms collect makes them particularly attractive targets for insiders who understand the commercial value of that data and have the access to take it.

Malware: Keyloggers, Rootkits, and Trojans

Beyond ransomware, the category of malicious software — malware — includes a range of tools that attackers use to gain and maintain unauthorized access, steal information, and disrupt operations. Understanding the principal types is part of the technical literacy that FinTech professionals need.

A keylogger is software that records every keystroke made on an infected device and transmits the captured data to the attacker. Keyloggers are used to steal passwords, account credentials, credit card numbers, and any other sensitive information typed on the compromised device. They can be installed through phishing links, malicious software downloads, or physical access to the device. Because they operate silently in the background and capture information at the point of entry — before it is encrypted in transit — they bypass many conventional security measures.

A rootkit is software designed to hide the presence of malware or unauthorized access from the operating system and security tools. A rootkit modifies the core functions of the operating system to conceal the attacker's presence, making the system appear to operate normally even while it is compromised. The term reflects the fact that these tools operate at the root — the deepest level — of the system. Rootkits are particularly dangerous because they can persist undetected for extended periods, giving attackers sustained access to systems and data.

A Trojan, or Trojan horse, is malware disguised as legitimate software. Like the Trojan horse of Greek mythology, it appears useful or benign but conceals malicious code that activates when installed. Trojans are commonly distributed through phishing emails, malicious downloads, and compromised software repositories. Banking Trojans are a specific variant designed to steal financial credentials — they may intercept banking app communications, inject false information into online banking sessions, or redirect transactions to attacker-controlled accounts.

Distributed Denial of Service Attacks

A distributed denial of service attack, or DDoS attack, overwhelms a target system with traffic from multiple sources — often a botnet of thousands of compromised devices — to make the system unavailable to legitimate users. Unlike attacks designed to steal data or maintain persistent access, DDoS attacks are primarily disruptive: their objective is to take systems offline, prevent legitimate users from accessing services, and impose operational and reputational costs on the target.

For FinTech firms, DDoS attacks against payment processing infrastructure, trading platforms, and customer-facing applications can have immediate operational and financial consequences. A payment processing platform that is offline for even a few hours during peak transaction periods loses revenue on every transaction it cannot process, faces customer service demands, and risks losing merchant relationships. Trading platforms that experience outages during volatile market periods face regulatory scrutiny and customer claims. The operational impact of successful DDoS attacks has driven significant investment in DDoS mitigation infrastructure across the financial services sector.

Hacktivism

Hacktivism is the use of hacking to promote a political cause. Hacktivists — individuals and groups who combine technical attack capabilities with political motivation — target organizations they associate with policies, practices, or values they oppose. Their objectives may include disrupting operations to impose costs, exposing internal information to generate embarrassment or public accountability, defacing public-facing websites to spread political messages, or demonstrating vulnerability to attract attention to a cause.

The distinction between hacktivism and other forms of cybercrime lies primarily in motivation: hacktivists are not primarily seeking financial gain. They are seeking political impact. The group Anonymous, which has conducted high-profile hacktivist operations against financial institutions, government agencies, and corporations since the mid-2000s, exemplifies the category — its operations have ranged from DDoS attacks on PayPal and Visa during the WikiLeaks controversy to the exposure of internal documents from organizations it targeted for political reasons.

For FinTech firms, hacktivism is a threat that may be difficult to anticipate through conventional threat intelligence because it is driven by political and reputational factors rather than the financial calculations that motivate criminal attackers. A FinTech firm that becomes associated with a controversial policy position, a high-profile enforcement action, or a politically charged product decision may find itself targeted by hacktivists who have no prior interest in the firm. The appropriate response — both defensive and reputational — requires understanding hacktivism as a distinct category with distinct characteristics.

Nation-State Threats

Nation-state cyber operations represent the most sophisticated end of the threat spectrum. Governments with significant intelligence and military capabilities — including Russia, China, North Korea, and Iran, among others — conduct cyber operations against financial infrastructure for multiple purposes: intelligence collection, economic espionage, pre-positioning for potential future disruption, and in some cases direct financial theft to fund sanctioned regimes.

North Korea's Lazarus Group has been one of the most active nation-state threat actors targeting the financial sector. The group has been attributed with attacks on the SWIFT interbank messaging network that resulted in the theft of approximately $81 million from Bangladesh's central bank in 2016, as well as attacks on cryptocurrency exchanges that have generated hundreds of millions of dollars for the North Korean government. These operations demonstrate that nation-state actors are not merely conducting espionage — they are conducting financially motivated attacks at a scale and sophistication level that no individual criminal group can match.

For most FinTech firms, nation-state attacks are not the primary operational threat — large financial market infrastructure and critical payment networks are more likely targets than individual FinTech startups. But the techniques and tools developed by nation-state actors migrate over time into the broader criminal ecosystem, and the interconnectedness of financial infrastructure means that a successful attack on one part of the system can have cascading effects on others.

Defenses: The Security Toolkit

Cybersecurity defense is not a single technology or technique — it is a layered architecture of controls designed to make successful attacks more difficult, detect intrusions more quickly, and limit the damage when breaches occur. The security professional's framework of defense in depth holds that no single control is sufficient and that multiple overlapping layers are required because any individual layer can be bypassed or defeated.

Firewalls

A firewall is a network security device — either hardware, software, or a combination — that monitors network traffic and filters out communications from untrusted sources based on defined security rules. Firewalls sit at the boundary between a trusted internal network and untrusted external networks, examining incoming and outgoing traffic and blocking traffic that does not meet the organization's security policies.

Modern firewalls have evolved significantly beyond the simple packet-filtering devices of the early internet era. Next-generation firewalls incorporate deep packet inspection, application-layer filtering, intrusion prevention capabilities, and integration with threat intelligence feeds that allow them to identify and block traffic associated with known malicious sources. A firewall that can identify and block traffic from a command-and-control server associated with a known malware family — rather than simply blocking traffic on a specific port — provides meaningfully stronger protection than older-generation devices.

For FinTech firms, firewall deployment must account for the distributed and API-driven nature of their architecture. A traditional perimeter firewall designed to protect a physical data center provides limited protection for a cloud-native FinTech firm whose infrastructure is distributed across multiple cloud providers and whose services communicate with hundreds of external APIs. Modern FinTech security architectures increasingly adopt a zero-trust model — in which no connection is assumed trusted regardless of its origin — rather than a perimeter model that assumes internal network traffic is safe.

Encryption

Encryption is the process of converting readable data into an unreadable format using a cryptographic algorithm, so that only parties with the appropriate decryption key can read it. It is one of the most fundamental and important controls in information security, and it operates at multiple layers of a FinTech firm's architecture.

Encryption in transit protects data as it moves between systems — between a customer's browser and a web server, between a mobile app and an API endpoint, between two data centers. Transport Layer Security, or TLS, is the standard protocol for encrypting web communications; the padlock icon in a browser address bar indicates an active TLS connection. Encryption in transit prevents attackers who intercept network traffic — through a public Wi-Fi network, a compromised network device, or a man-in-the-middle attack — from reading the captured data.

Encryption at rest protects data stored on servers, databases, and backup media. If an attacker gains access to a database containing customer financial information, encryption at rest means that the raw data files are unreadable without the corresponding decryption keys. This does not prevent a breach from occurring, but it can significantly limit the damage by preventing the attacker from reading the stolen data.

End-to-end encryption goes further — it ensures that data is encrypted from the point of creation to the point of reading, such that no intermediate party, including the service provider, can read the content.

End-to-end encryption is standard in messaging applications and is increasingly being applied to sensitive financial communications and document storage. For FinTech firms that handle particularly sensitive customer information, end-to-end encryption represents a meaningful privacy and security enhancement, though it also creates challenges for compliance monitoring and fraud detection that depend on the provider's ability to access transaction content.

Multi-Factor Authentication

Multi-factor authentication, or MFA — commonly referred to as two-factor authentication, or 2FA, when exactly two factors are required — requires users to verify their identity through more than one method before gaining access to a system. The three standard categories of authentication factors are: something you know (a password, a PIN, or a security question answer); something you have (a physical device such as a smartphone, a hardware token, or a smart card); and something you are (a biometric identifier such as a fingerprint, facial scan, or retinal pattern).

The rationale for multi-factor authentication is straightforward: a password alone is a single point of failure. Passwords can be guessed through brute-force attacks, stolen through phishing or keylogging, or exposed in data breaches at other services where the same password was reused. Adding a second factor means that an attacker who has obtained a user's password still cannot gain access without also having the second factor — which, in the case of a physical device, requires either stealing the device or compromising the channel through which the second factor is delivered.

Multi-factor authentication has meaningful limitations that FinTech professionals need to understand. SMS-based authentication — in which a one-time code is sent to the user's phone via text message — is vulnerable to SIM-swapping attacks, in which an attacker convinces a mobile carrier to transfer a victim's phone number to a SIM card the attacker controls. Once the attacker controls the phone number, they receive the SMS codes and can bypass 2FA. SIM-swapping attacks have been used to access cryptocurrency wallets, brokerage accounts, and banking apps with significant financial consequences.

Biometric authentication introduces a different category of limitation. A stolen password can be changed; a compromised fingerprint or facial scan cannot. If biometric authentication data is stored insecurely and exposed in a breach, the affected users cannot simply update their biometrics the way they would update a password. The immutability of biometric identifiers makes the security of the systems that store and process them especially critical. And biometric systems can be vulnerable to spoofing — high-quality photographs have been used to defeat some facial recognition systems, and silicone fingerprint replicas have been used against some fingerprint scanners.

Authenticator apps — which generate time-based one-time passwords that expire after thirty seconds — provide stronger protection than SMS-based 2FA because the codes are generated locally on the device rather than transmitted over a network channel that can be intercepted or redirected. Hardware security keys, such as those based on the FIDO2 standard, provide the strongest commonly available form of multi-factor authentication because they require physical possession of the key and are resistant to phishing attacks. The trade-off is between the security strength of the authentication method and the friction it imposes on users — a friction that affects adoption rates and user experience in ways that have commercial implications for FinTech platforms.

Access Controls

Access control is the practice of limiting access to systems, data, and functionality to those who have a legitimate need for it. The principle of least privilege holds that every user, process, and system should have access only to the resources required for its specific function — no more. Implementing least privilege means that a customer service representative who needs to view account balances does not also have access to modify transaction records, and an application server that needs to read from a database does not also have write access.

The benefit of least privilege is that it contains the damage from any individual compromise. If a customer service representative's credentials are stolen in a phishing attack, the attacker gains the limited access that representative had — not the full access of a system administrator. If a web application is compromised, the attacker can only access the data and functions that the application was authorized to use — not the entire database.

Implementing tight access controls has real costs, and FinTech professionals need to understand the trade-off. Maintaining granular access controls is time-consuming, expensive and difficult to do, and tough to automate — particularly in organizations that are growing quickly and changing their systems frequently. The operational burden of access control — revoking access when employees change roles, reviewing permissions regularly, managing the complexity of many different access levels — is substantial. This tension between security rigor and operational efficiency is one of the recurring challenges of cybersecurity practice.

Honey Pots

A honey pot is a security resource — a simulated system, server, or network environment — designed to look attractive to attackers but that is actually closely monitored, so that everything an attacker does within it is recorded and studied. Honey pots serve a deception and intelligence function: by presenting a convincing target that contains no real production data or critical systems, they attract attackers who believe they have found a valuable resource. The attackers' activity — the techniques they use, the vulnerabilities they attempt to exploit, the data they search for, and the tools they deploy — is captured and analyzed.

It is important to be precise about what honey pots are: they are a defensive security tool, not an attack tool. Keyloggers, rootkits, and Trojans are programs used by attackers to gain unauthorized access or steal information. Honey pots are used by defenders to detect intrusions, study attacker behavior, and gain intelligence about the tools and techniques in current use. A FinTech firm that deploys a honey pot configured to resemble a database containing customer payment information may attract attackers who have already breached the perimeter — and the honey pot's monitoring will reveal the intrusion and the attacker's methods before any real data is accessed.

Honey pots also serve an early warning function. An attacker who begins probing a honey pot has often already gained some access to the network — meaning that the honey pot's detection capability triggers an alert at a stage of the attack before critical systems have been reached. This early warning value is one of the primary reasons security teams deploy honey pots within internal networks rather than just at the perimeter.

White Hat Hackers and Ethical Hacking

Not every skilled hacker became a criminal. As the hacking community matured through the 1980s and 1990s, a distinct professional track emerged for individuals who applied the same technical skills toward defense rather than attack. These practitioners came to be called white hat hackers — a term drawn from the old Western film convention of dressing heroes in white and villains in black. The distinction is not about capability. It is about authorization and intent.

A white hat hacker is a security professional who attempts to breach systems, networks, or applications with the explicit permission of the organization that owns them. The goal is to find vulnerabilities before malicious actors do. This practice is formally known as penetration testing, or pen testing, and it has become a standard component of enterprise security programs across the financial services industry. A FinTech firm that commissions a penetration test is essentially hiring an expert to attack its own systems under controlled conditions and report back on every weakness found.

The value of white hat work is that it tests defenses against realistic attack techniques rather than theoretical ones. Automated scanning tools can identify known vulnerabilities in software configurations, but they cannot replicate the judgment, creativity, and persistence of a skilled human attacker. A white hat hacker attempting to breach a payment platform will chain together multiple small weaknesses — a misconfigured API endpoint, an overly permissive employee account, a gap in monitoring coverage — in ways that automated tools miss. That is precisely how real attackers operate.

White hat hackers operate under clearly defined rules of engagement: a written agreement that specifies what systems can be tested, what methods are permitted, when testing will occur, and how findings will be reported. This authorization is what separates their work from criminal intrusion. The Computer Fraud and Abuse Act, which criminalizes unauthorized access to computer systems, does not apply to authorized penetration testing — but the authorization must be explicit and documented. A security professional who exceeds the scope of a penetration test agreement, accessing systems not covered by the engagement, has crossed into legally and professionally prohibited territory regardless of their intent.

Beyond penetration testing, white hat hackers contribute to security through bug bounty programs — formalized initiatives in which organizations invite independent researchers to identify and responsibly disclose vulnerabilities in exchange for financial rewards. Major FinTech firms including PayPal, Square, and Coinbase have operated bug bounty programs that have surfaced significant vulnerabilities before malicious actors could exploit them. This model extends the white hat community beyond full-time security employees to a broader global pool of independent researchers, each with different technical specializations and perspectives.

The distinction between white, gray, and black hat hackers reflects authorization status, not technical skill. Black hat hackers operate without authorization for malicious or personal gain. Gray hat hackers occupy an ambiguous middle ground — they may access systems without permission but disclose vulnerabilities rather than exploit them, a practice that remains legally risky regardless of intent. White hat hackers operate with explicit authorization, within defined boundaries, in service of the organization's security. For FinTech professionals, the practical takeaway is straightforward: hiring skilled adversaries to find your weaknesses on your terms is far preferable to having unskilled or malicious ones find them on theirs.

Security Monitoring and Threat Intelligence

Beyond specific technical controls, effective cybersecurity requires continuous monitoring — the real-time collection and analysis of security events across an organization's systems to identify suspicious activity, detect intrusions, and trigger responses. Security information and event management systems, or SIEM systems, aggregate log data from across an organization's infrastructure — firewalls, servers, applications, endpoints — and apply correlation rules and analytics to identify patterns that indicate potential security incidents.

Threat intelligence is the practice of collecting, analyzing, and acting on information about the threat landscape — the actors, tools, techniques, and vulnerabilities that are currently being actively exploited. FinTech firms subscribe to threat intelligence feeds that provide real-time information about newly identified malware, newly discovered vulnerabilities, indicators of compromise associated with known threat actors, and emerging attack techniques. This intelligence allows security teams to proactively update their defenses before known threats reach their systems rather than waiting to detect an intrusion after it has occurred.

FinTech-Specific Vulnerabilities

While the threats and defenses described above apply broadly across the technology sector, FinTech firms face a specific set of vulnerabilities that are products of their particular operating model. Understanding these vulnerabilities is essential for designing security programs that address the actual risk profile of a FinTech business rather than a generic technology company.

The API Attack Surface

FinTech firms are defined by their connectivity. They connect to banks, payment networks, credit bureaus, data providers, and each other through application programming interfaces — APIs. The Section 1033 open banking framework discussed in Chapter 8 will expand this connectivity further, making API-based data sharing a standard feature of the financial services landscape. Every API connection is a potential attack vector.

API vulnerabilities are among the most actively exploited in the current threat environment. Poorly authenticated APIs can be accessed without valid credentials. APIs that return more data than necessary — a vulnerability called excessive data exposure — provide attackers with information they should not have. APIs that do not properly validate inputs can be manipulated through injection attacks that cause them to execute unintended actions. And APIs that lack rate limiting can be subjected to automated attacks that systematically enumerate account information or attempt credential stuffing.

For FinTech firms building on open banking infrastructure, the API security challenge extends beyond their own systems to their partners'. A well-secured FinTech firm that connects to a partner with poor API security may find that the partner's vulnerability becomes its own exposure. The 2019 Capital One data breach — in which approximately 100 million customer records were accessed — exploited a misconfigured web application firewall in Capital One's cloud environment, illustrating how configuration errors in complex, interconnected architectures can create major exposures.

Third-Party and Supply Chain Risk

FinTech firms depend on extensive ecosystems of third-party vendors, cloud providers, software libraries, and service providers. Each dependency introduces potential security risk. A vulnerability in a widely used software library can simultaneously expose every FinTech firm that includes it in their codebase. A compromise of a cloud service provider can affect thousands of customers at once. A security failure at a payment processor, identity verification provider, or data feed vendor can cascade through all the FinTechs that rely on it.

The 2020 SolarWinds attack — in which attackers compromised the software update mechanism for SolarWinds' Orion network management software, inserting a backdoor that was then distributed to approximately 18,000 organizations through routine software updates — illustrated the scale and subtlety of supply chain attacks. The targets included financial institutions, government agencies, and technology companies that had done nothing wrong themselves: they had simply installed a software update from a trusted vendor. For FinTech firms, supply chain risk management — evaluating the security posture of vendors, monitoring for vendor-related indicators of compromise, and maintaining contingency plans for vendor failures — is a significant and often underresourced security function.

Rapid Scaling and Security Debt

FinTech startups characteristically prioritize speed to market, user experience, and growth over security infrastructure. In the early stages of a FinTech company, engineering resources are consumed by product development, security controls that slow down user onboarding or complicate the development process may be deliberately deferred, and the small team size means there may be no dedicated security function at all. This approach creates security debt — the accumulated technical and process vulnerabilities that result from underprioritizing security during growth.

Security debt is not merely a startup problem. It persists as firms scale, because the systems built without adequate security controls require significant investment to retrofit — an investment that must compete with product roadmap priorities. A payment platform that processes millions of transactions per day on infrastructure built to move fast in year one faces a complex and expensive security remediation challenge in year five. The security assumptions baked into early architectural decisions can be very difficult to change without rebuilding significant parts of the system.

This is one reason why the GLBA Safeguards Rule's requirement for a written information security program is particularly meaningful for FinTech firms — it establishes a floor of security planning that encourages firms to address security systematically rather than reactively. And it is why investors and acquirers increasingly conduct detailed security due diligence on FinTech firms: the cost of inheriting security debt from an acquisition can far exceed the cost of the security improvements that were deferred during the company's growth phase.

Mobile and Remote Attack Surface

Most FinTech products are delivered primarily through mobile applications. Mobile platforms introduce security challenges that differ from traditional web or desktop applications. Mobile apps are distributed through app stores that, despite review processes, have hosted malicious applications that impersonate legitimate financial services. Mobile devices are frequently used on untrusted networks — public Wi-Fi, cellular networks in foreign countries — that may be subject to interception. Mobile devices are lost and stolen at significant rates, creating risks of unauthorized physical access. And the mobile ecosystem's fragmentation — the range of operating system versions, device types, and configurations that a FinTech app must support — makes consistent security testing and patching challenging.

The remote work environment that became standard during and after the COVID-19 pandemic has extended the attack surface further. Employees accessing FinTech systems from home networks, personal devices, and varied physical environments present security management challenges that were not part of the traditional enterprise security model. The assumption that all legitimate access originates from a controlled corporate network — an assumption that traditional security architectures often built on — no longer holds for most FinTech firms.

Incident Response: What to Do Under Cyberattack

No cybersecurity program is impenetrable. The question for FinTech firms is not whether they will face a significant cybersecurity incident but how they will respond when they do. A well-designed incident response program can mean the difference between a contained disruption and a catastrophic breach — and between a firm that emerges from an incident with its reputation and relationships intact and one that faces regulatory action, civil litigation, and customer defection.

When a firm is under cyberattack, the response involves several interconnected steps that must be executed simultaneously and systematically. Looking at the impact means assessing what has been affected — which systems, which data, which operational capabilities — and understanding the scale and severity of the disruption. A ransomware attack that has encrypted the firm's development environment is a different situation from one that has encrypted the core transaction processing system, and the response must be calibrated to the actual impact.

Identifying the source, cause, and intention of the attack is critical for effective response. A DDoS attack requires different mitigation than a data breach. A financially motivated criminal operation requires a different response than a nation-state intrusion that may have established persistent access. Understanding whether the attacker's objective was disruption, theft, espionage, or ransom informs both the immediate technical response and the longer-term remediation strategy. Forensic investigation — preserving evidence, analyzing logs, tracing the attacker's path through the system — is a specialized discipline that many FinTech firms retain on a retainer basis with incident response firms precisely because the capacity to conduct it is needed immediately when an incident occurs.

Reporting the incident to victims and law enforcement is both a legal obligation and a practical necessity. Firms that have experienced data breaches affecting customer information have notification obligations under state breach notification laws, the GLBA Safeguards Rule, and potentially other sectoral regulations. Customers whose data has been compromised need to know so they can take protective measures — monitoring their accounts, placing fraud alerts on their credit files, and changing credentials. Law enforcement reporting — to the FBI's Internet Crime Complaint Center, to relevant sector-specific agencies — both fulfills regulatory obligations and contributes to the collective intelligence about the attack that may help protect other potential victims.

Looking at legal liabilities means assessing the firm's exposure from the incident — its obligations to customers, its potential regulatory liability, its contractual obligations to partners, and its obligations under cyber insurance policies. Legal counsel should be engaged early in the incident response process, both to manage the legal dimensions and to ensure that the firm's communications and actions during the incident do not inadvertently waive privileges or create admissions that complicate subsequent litigation or regulatory proceedings.

Underlying all of these steps is the need for a prepared, documented, and practiced incident response plan. Firms that are attempting to figure out who does what, who has authority to make decisions, and how to communicate with customers and regulators in the middle of an active attack are at a severe disadvantage compared to firms that have rehearsed these decisions in advance. The NIST Cybersecurity Framework and other industry standards provide structured templates for incident response planning, and the GLBA Safeguards Rule requires financial institutions to have an incident response plan that is periodically tested.

The Human Element: Security Culture and Training

Technical controls are necessary but not sufficient for effective cybersecurity. Because many of the most common and most damaging attacks — phishing, social engineering, insider threats — exploit human behavior rather than technical vulnerabilities, the human element of security is as important as the technical architecture. Building a security-aware organizational culture is not a soft consideration alongside the hard technical work of deploying firewalls and encryption — it is a core component of an effective security program.

Security awareness training — educating employees about the threats they face, the techniques attackers use, and the behaviors that reduce risk — is a foundational security control. Training that teaches employees to recognize phishing emails, to be skeptical of unusual requests for credentials or data transfers, and to report suspicious activity to the security team is directly correlated with reduced susceptibility to the attacks that training addresses. Simulated phishing campaigns — in which the security team sends fake phishing emails to employees and tracks who clicks — are a standard tool for both measuring awareness and reinforcing training through immediate feedback.

Security culture extends beyond formal training to the values and norms that govern how employees think about and relate to security in their daily work. A culture in which security is viewed as an obstacle to be worked around — in which employees share passwords because it is more convenient, where shadow IT flourishes because the official tools are too cumbersome, and where security concerns are not raised because they will slow down product releases — is a culture that will produce security failures regardless of the technical controls in place. Building a culture in which security is a shared professional value requires leadership commitment, consistent messaging, and a security function that is seen as a partner to the business rather than an impediment.

Key Takeaways

Cybersecurity is a fundamental operational requirement for FinTech firms, not a peripheral technical concern. FinTech platforms are high-value targets because of the financial data they hold, the payment infrastructure they operate, and the digital-first architecture that creates a large and complex attack surface.

The history of hacking evolved from its origins in the 1960s, when a hacker was a creative programmer who wrote very elegant or clever programs, through the phone phreaking era of the 1970s and 1980s, through the commercialization of the internet and the emergence of organized cybercrime in the 1990s and 2000s, to the current threat landscape characterized by three overlapping actor categories: financially motivated criminal organizations, nation-state actors, and hacktivists. Hacktivism specifically is the use of hacking to promote a political cause — distinct from financially motivated crime or state-sponsored espionage.

The major threat categories facing FinTech firms include phishing and spear phishing, which exploit human psychology to steal credentials and initiate unauthorized transactions; ransomware, which encrypts victim data and demands payment for its return; social engineering, which manipulates people rather than exploiting code; and insider threats, which are a major problem specifically because internal actors are already authorized and have privileges within the system, making their malicious activity harder to detect than external intrusions. Attack tools used by adversaries include keyloggers, which record keystrokes; rootkits, which conceal malware by operating at the system's core; and Trojans, which disguise malicious code as legitimate software.

Core defensive controls include firewalls, which monitor and filter communications from untrusted sources; encryption in transit and at rest, which protects data from interception and unauthorized access; multi-factor authentication, which requires verification through multiple factors and reduces the risk of credential theft; and access controls implementing the principle of least privilege. Honey pots are websites or simulated systems that look attractive to hackers but are closely monitored so that everything an attacker does is recorded and studied — they are a defensive intelligence tool, not an attack tool.

Multi-factor authentication has important limitations. SMS-based 2FA is vulnerable to SIM-swapping attacks. Biometric authentication cannot be reset if compromised. Authenticator apps and hardware security keys provide stronger protection but impose greater friction on users. The security-versus-convenience trade-off in authentication design is one of the central practical tensions in FinTech security.

FinTech-specific vulnerabilities include the broad API attack surface created by open banking and digital-first architectures, third-party and supply chain risk from extensive vendor ecosystems, security debt accumulated during rapid growth phases, and the mobile and remote access attack surface. These vulnerabilities are products of the same architectural choices that give FinTech firms their competitive advantages.

When a firm is under cyberattack, the response requires simultaneously: looking at the impact; identifying the source, cause, and intention of the attack; reporting to victims and law enforcement; and looking at legal liabilities. An effective response depends on preparation — a documented, tested incident response plan that has been rehearsed before the attack occurs, not assembled during it.

Chapter 10 builds on this technical foundation to examine the legal and policy landscape governing cybersecurity obligations, with particular attention to the Cybersecurity Information Sharing Act of 2015 and the ongoing debate about the appropriate roles of government and the private sector in national cybersecurity.

Chapter 10: Cybersecurity Law and Policy — CISA and the Regulatory Response

Introduction

Chapter 9 established the threat landscape that FinTech firms navigate: sophisticated adversaries using phishing, ransomware, insider access, and nation-state capabilities to attack systems that hold some of the most sensitive and valuable data in the economy. The chapter closed with a question implicit in the incident response framework — what happens after the attack? A firm that has successfully contained a breach, notified affected customers, and engaged law enforcement has done the right things operationally. But it has also been the sole actor in a defense that might have been more effective if others had known what was coming.

That is the problem cybersecurity law and policy is designed to address. No single FinTech firm, however well-resourced its security program, can see the full threat landscape on its own. Attackers share tools, techniques, and intelligence freely within their communities. Defenders, constrained by competitive concerns and legal uncertainty, have historically been reluctant to share what they know with each other or with government. The result is a structural asymmetry: adversaries operate collectively while defenders operate in isolation.

The Cybersecurity Information Sharing Act of 2015, known as CISA, was Congress's primary legislative response to this asymmetry. Enacted in December 2015 after years of debate, CISA created a framework for voluntary sharing of cyber threat information between private companies and the federal government, with liability protections for participating firms and privacy safeguards designed to limit the personal information that flows through the sharing process. It was designed to remove the barriers — legal risk, competitive concern, and institutional reluctance — that had prevented effective information sharing and to create the infrastructure for collective defense.

CISA is also the subject of the Module 5 debate assignment, in which you will argue in favor of the law against an AI system arguing against it. That assignment requires you to understand CISA deeply enough to make the strongest possible case for it while also anticipating and rebutting the strongest arguments against it. This chapter is written with that purpose in mind. It presents both sides of the CISA debate with care, because preparing to argue one side of a complex policy question requires genuinely understanding both.

Beyond CISA, this chapter covers the broader legal framework governing cybersecurity obligations for FinTech firms — the Computer Fraud and Abuse Act, sector-specific regulatory requirements, the NIST Cybersecurity Framework, and the respective roles of government and the private sector in national cybersecurity. It closes with the tension that runs through all cybersecurity policy: how to balance the security benefits of surveillance and information sharing with the civil liberties concerns those activities inevitably raise.

The Collective Defense Problem

To understand why CISA was created, you need to understand the problem it was trying to solve. Cybersecurity in the financial sector — as in every sector — has historically been organized around the individual firm. Each bank, each payment processor, each FinTech platform builds its own security program, deploys its own controls, analyzes its own threat data, and responds to its own incidents. The knowledge generated by each firm's security experience — what attacks were attempted, what techniques were used, what vulnerabilities were exploited — largely stays within that firm.

This isolation is problematic for a specific reason: the threat landscape is shared. When a criminal organization develops a new phishing kit targeting financial institutions, it uses that kit against multiple firms in sequence. When a nation-state actor identifies a vulnerability in widely used financial software, it can exploit that vulnerability against every financial firm running that software. When a ransomware variant spreads through financial sector networks, it does so across organizational boundaries. The firms that are attacked first may develop defenses and detect the attack. The firms that are attacked second, third, and fourth — which might have been protected if they had known what was coming — are left to discover the threat on their own.

The contrast with the attacker community is stark. Criminal forums, dark web marketplaces, and hacker networks operate as information-sharing ecosystems in which new tools, techniques, and vulnerabilities are traded, sold, and discussed openly. A ransomware variant that one criminal group develops may be licensed to other groups within weeks. A phishing kit targeting a specific bank may be available for purchase by any sufficiently motivated attacker. The attackers' collective intelligence is a genuine operational advantage over defenders who cannot or do not share what they know.

Information sharing initiatives existed in the financial sector before CISA. The Financial Services Information Sharing and Analysis Center, known as FS-ISAC, was founded in 1999 specifically to facilitate threat intelligence sharing among financial institutions. FS-ISAC has been one of the more successful sector information sharing organizations and is regularly cited as evidence that voluntary sharing can work. But CISA's proponents argued that FS-ISAC and similar initiatives were insufficient: participation was limited, sharing was often delayed rather than real-time, firms were reluctant to share sensitive information about incidents they had experienced due to concerns about legal liability and competitive exposure, and the channel between industry and government for mutual sharing of threat intelligence was poorly developed.

The legal liability concern was particularly significant. A firm that shared information about a security incident with a competitor or with a government agency faced potential exposure to claims that it had disclosed confidential customer information, violated privacy laws, or provided information that somehow harmed a third party. Even if such claims were unlikely to succeed, the threat of litigation created a powerful incentive for lawyers to advise against sharing. CISA was designed in substantial part to eliminate that incentive by providing a legal safe harbor for firms that shared threat information in accordance with the statute's requirements.

The Cybersecurity Information Sharing Act: Purpose and Provisions

The Cybersecurity Information Sharing Act of 2015 was enacted as part of the Consolidated Appropriations Act of 2016, signed into law on December 18, 2015. Its stated purpose is to improve cybersecurity in the United States through enhanced sharing of information about cybersecurity threats, and to provide certain protections to entities that voluntarily share or receive cyber threat indicators and defensive measures.

Core Definitions

CISA operates around two central concepts: cyber threat indicators and defensive measures. A cyber threat indicator is information that is necessary to describe or identify a malicious reconnaissance activity; a method of defeating a security control; a security vulnerability; a method of causing a user to unwillingly enable the execution of code; malicious cyber command and control; the actual or potential harm caused by an incident; or any other attribute of a cybersecurity threat. In practical terms, a cyber threat indicator might be an IP address known to be associated with a malware command-and-control server, a domain name used in a phishing campaign, a hash value identifying a malicious file, or a pattern of network traffic associated with an intrusion.

A defensive measure is an action, device, procedure, signature, technique, or other measure applied to an information system or information that is stored on, processed by, or transiting an information system that detects, prevents, or mitigates a known or suspected cybersecurity threat or security vulnerability. Defensive measures are the practical countermeasures — firewall rules, intrusion detection signatures, software patches — that firms deploy in response to threat intelligence.

The Sharing Mechanism

CISA authorizes private entities to monitor their own information systems, as well as information systems operated on their behalf, for cybersecurity purposes. It authorizes the sharing of cyber threat indicators and defensive measures with other private entities and with the federal government. Critically, sharing under CISA is entirely voluntary — no firm is required to share anything, and choosing not to participate carries no legal consequence.

For sharing with the federal government, CISA designated the Department of Homeland Security as the primary civilian receiving agency. DHS was required to establish a capability and process for real-time automated sharing of unclassified cyber threat indicators in the federal government's possession with private entities. This Automated Indicator Sharing capability, known as AIS, allows participating firms to receive threat indicators directly from the government as well as to submit their own — creating a two-way information channel rather than a one-directional reporting requirement.

The routing of government-facing sharing through DHS was a deliberate architectural choice. Critics of earlier cybersecurity information sharing proposals had objected to routing sharing through the National Security

Agency or military channels — institutions whose primary missions involve intelligence collection and whose involvement in civilian cybersecurity raised concerns about government surveillance and militarization of domestic security. CISA's civilian DHS-centered architecture was designed to address those concerns, though critics argued it did not fully resolve them.

Privacy Protections and Personal Information Scrubbing

CISA includes privacy protections designed to limit the amount of personal information that flows through the sharing system. Before sharing a cyber threat indicator with any other entity — private or government — a firm must review the indicator and remove any information the firm knows to be personal information not directly related to a cybersecurity threat. This scrubbing obligation applies to both private-to-private and private-to-government sharing.

When DHS receives shared information from private entities, it must implement a process to review and scrub additional personal information before the shared data is disseminated to other government agencies. The law requires this scrubbing to be conducted in an automated manner in near-real time — meaning that the process is designed to move quickly enough to support the real-time sharing capability the law is trying to create, rather than introducing delays that would reduce the intelligence value of shared indicators.

The law also includes use limitations on how shared information can be used by the government. Shared cyber threat indicators and defensive measures may be used for cybersecurity purposes, the identification and prosecution of cybersecurity crimes, the protection of individuals from the danger of death or serious bodily harm, and a handful of other specified national security and law enforcement purposes. They may not be used to regulate the lawful activities of a private entity or to conduct surveillance of U.S. persons for purposes unrelated to cybersecurity.

Liability Protections

The liability protection provisions are central to CISA's design. A private entity that monitors information systems or shares cyber threat indicators and defensive measures in accordance with CISA is not liable under any federal or state law for such monitoring and sharing. This protection applies to claims that the sharing violated the Computer Fraud and Abuse Act, the Electronic Communications Privacy Act, the Stored Communications Act, antitrust law, or any other federal law. It also applies to any state law claim arising from the sharing.

The liability protection has three important limitations. First, it applies only to sharing that meets CISA's requirements — the scrubbing of personal information must have been conducted, the sharing must have been for a cybersecurity purpose, and the recipient must have been an authorized entity. Sharing that does not comply with CISA's requirements does not receive the protection. Second, it does not protect sharing that a firm knows to include information that is false. Third, it does not protect against government enforcement actions for violations of other laws that are not specifically listed in the statute.

From the perspective of a FinTech firm's legal counsel, the liability protection substantially changes the risk calculus around sharing. Before CISA, sharing information about a security incident carried legal risk that was difficult to quantify. After CISA, a firm that follows the statute's requirements can share with a reasonable degree of confidence that the sharing itself will not generate legal exposure. That shift in the risk calculation was precisely what CISA's proponents argued was necessary to make voluntary sharing work at scale.

Oversight and Accountability Mechanisms

CISA includes several oversight mechanisms intended to ensure accountability in how shared information is handled. DHS and the Department of Justice were required to develop privacy and civil liberties guidelines governing the receipt, retention, use, and disclosure of cyber threat indicators. These guidelines were required to be developed in consultation with the Privacy and Civil Liberties Oversight Board — an independent executive branch agency established to oversee government surveillance programs — and to be made publicly available.

The law also requires annual reports to Congress on the sharing program's activities, including the number of cyber threat indicators received, the number scrubbed for personal information, and an assessment of the program's effectiveness. The Inspector General of the Intelligence Community and the Inspector General of the Department of Homeland Security are required to submit reports on the implementation of the law's privacy protections. These reporting requirements are designed to provide Congress and the public with visibility into how the program is operating without compromising the operational security of specific shared threat intelligence.

The Case For CISA

Understanding the strongest version of the argument in favor of CISA is essential for engaging seriously with a genuine and consequential policy question. The pro-CISA argument rests on several interconnected claims about the nature of the cybersecurity threat, the limits of individual firm defense, and the specific mechanisms CISA uses to address them.

The Collective Defense Argument

The most fundamental argument for CISA is structural: the cybersecurity threat to financial infrastructure is a collective action problem that individual firm-level defense cannot solve. When a sophisticated attacker — a nation-state actor, an organized criminal network, or a highly capable hacktivist group — develops a new technique, every firm in the sector is a potential target. The first firm attacked has no way of knowing it is facing a new threat; it discovers that fact through the attack itself. If that firm can share what it learned — the indicators of compromise, the attack signatures, the vulnerability exploited — in near real time with every other firm in the sector, the subsequent targets can implement defenses before they are attacked.

This collective defense benefit is real and measurable. The financial sector's experience with FS-ISAC illustrates it: member institutions that receive threat intelligence through the sharing network can deploy defensive countermeasures against known threats before those threats reach their own systems. A bank that receives an indicator that a specific IP address is actively used in a targeted financial sector attack can block that IP address before a single packet from it reaches the bank's network. Without the sharing, the bank would not know to block it until the attack had already begun.

CISA's contribution to this collective defense model is twofold. It provides the liability protection that removes the legal barrier to participation. And it establishes the government-to-private sector channel that allows threat intelligence from classified and law enforcement sources — information about nation-state actors, criminal organizations, and vulnerabilities identified through government intelligence — to flow to private sector defenders who can use it without requiring security clearances.

The Liability Shield as an Enabling Mechanism

Before CISA, the practical obstacle to information sharing was not primarily a lack of willingness — it was legal risk. Companies that had experienced security incidents were reluctant to share details because disclosure could be interpreted as an admission of negligence, could generate claims from third parties whose data was involved, and could potentially violate the privacy laws governing the information the attacker had accessed. Legal counsel routinely advised against sharing for these reasons, even when the security benefits of sharing were apparent.

CISA's liability shield was designed specifically to solve this problem. By providing explicit federal protection for sharing that complies with the statute's requirements, CISA changed the legal environment in which sharing decisions are made. A FinTech firm that shares a threat indicator after scrubbing personal information and routing through DHS can do so with the assurance that the sharing itself will not generate a lawsuit or a regulatory enforcement action. That assurance changes the calculus from one in which sharing is legally risky to one in which not sharing is the unprotected choice.

Leveling the Playing Field

A significant benefit of CISA's information sharing framework that its proponents emphasize is its effect on smaller firms. A large bank with hundreds of security analysts and a mature threat intelligence program can monitor the global threat landscape, track adversary activity, and develop its own indicators of compromise. A small FinTech startup with a team of three security engineers cannot do the same. Without information sharing, the small firm's threat intelligence capability is a fraction of the large firm's, and it faces the same adversaries at a severe disadvantage.

The AIS platform and the sector-specific information sharing organizations that CISA supports level this playing field to a degree. When large financial institutions share threat indicators through FS-ISAC or directly with DHS and those indicators are redistributed through AIS, smaller firms receive threat intelligence that they could not have developed independently. The collective intelligence of the entire sector becomes available to every participant, regardless of the size of their individual security program. CISA's proponents argue that this democratization of threat intelligence is one of its most significant practical benefits.

The National Security Dimension

For threat categories that extend beyond conventional criminal activity — nation-state attacks on financial market infrastructure, for example, or coordinated attacks designed to disrupt payment systems as an instrument of geopolitical pressure — individual firm defense is structurally inadequate. No private company can counter a nation-state adversary through its own security program alone. These threats require government capabilities: intelligence about adversary intentions and capabilities, diplomatic and law enforcement tools to deter and prosecute, and potentially military responses in the most extreme cases.

CISA's proponents argue that the law creates the information infrastructure necessary for government to apply its unique capabilities to the private sector's threat environment. When a FinTech firm shares threat indicators that enable the government to attribute an attack to a specific nation-state actor, that attribution can support diplomatic, economic, and potentially military responses that are available to the government but not to the private sector. The government's ability to respond to cyber threats against critical financial infrastructure depends in part on having timely, detailed information about what those threats look like — and CISA creates a legal mechanism for that information to flow from those who see the attacks to those who have the tools to respond to them.

The Case Against CISA

The civil liberties and privacy communities have raised sustained objections to CISA since it was introduced — objections that the law's proponents have responded to but that have not been resolved to critics' satisfaction. Understanding these objections in their strongest form is essential for a credible debate, because an argument that cannot anticipate and address the opposition's best points is not a fully developed argument.

The Privacy Objection: Scrubbing Is Insufficient

The most fundamental objection to CISA is that the privacy protections built into the law are inadequate to prevent the government from acquiring substantial personal information about American citizens through the sharing mechanism. Critics argue that cyber threat indicators — even after the personal information scrubbing the law requires — regularly contain information that identifies individuals: IP addresses that can be traced to specific people, email addresses, behavioral patterns that are effectively unique to individual users, and network metadata that reveals who was communicating with whom.

The scrubbing obligation requires firms to remove information they know to be personal information not directly related to a cybersecurity threat. Critics argue that this standard is too permissive — that the category of personal information directly related to a cybersecurity threat is broad enough to include most of the personal information that appears in threat indicators, and that the scrubbing process as implemented is unlikely to catch everything. The Electronic Frontier Foundation, the American Civil Liberties Union, and other civil liberties organizations argued at the time of CISA's passage that the privacy protections were window dressing — meaningful enough to give the law political cover but insufficient to prevent significant personal data from flowing to government agencies.

The Surveillance Expansion Argument

A related and broader objection is that CISA, whatever its stated cybersecurity purpose, functions primarily as a surveillance expansion measure. The law allows private companies to share information with the government under a cybersecurity label with limited restrictions on how the government uses that information thereafter. The use limitations in the law — restricting use to cybersecurity, specific law enforcement, and national security purposes — are, critics argue, broad enough to encompass nearly any government interest and narrow enough to provide no meaningful constraint.

This concern is not purely hypothetical. The history of government surveillance programs in the United States includes multiple examples of information collected for one stated purpose being used for broader intelligence gathering. The revelations about NSA surveillance programs made public by Edward Snowden in 2013, which included the collection of metadata on virtually all domestic phone calls under a broad interpretation of the PATRIOT Act, were fresh in Congress's memory during CISA's debate. Civil liberties advocates argued that CISA's structure was analogous — a broad legal authority with weak oversight mechanisms that could be used to justify surveillance far beyond the law's stated purpose.

The Effectiveness Question

Critics also questioned whether CISA would actually improve cybersecurity, arguing that the law addresses a less important barrier than its proponents claimed. Most of the threat indicators that firms would be likely to share, critics argued, are low-value — IP addresses that change frequently, domain names that are abandoned after use, malware hashes that evolve rapidly. By the time shared indicators are processed and distributed, sophisticated attackers have already moved on to new infrastructure. The threat actors most dangerous to financial infrastructure — nation-state groups with significant operational security discipline — are precisely the adversaries against whom shared indicators are least useful.

The effectiveness critique also questioned whether the liability protection was actually the binding constraint on sharing. Some research suggested that firms' reluctance to share was driven more by competitive concerns — not wanting to disclose that they had been attacked — and by the operational burden of processing shared information than by legal liability. If so, CISA's primary mechanism — the liability shield — addressed the wrong problem, and the sharing it enabled would not meaningfully improve collective defense against the most serious threats.

The Scope Creep Concern

A third category of objection focuses on the potential for CISA's authorities to expand over time in ways that Congress did not intend. The law's definitions of cyber threat indicator and defensive measure are broad, and the use exceptions that allow shared data to be used for purposes beyond cybersecurity are written in terms general enough to accommodate expansive interpretation. Civil liberties advocates worry that as the information sharing infrastructure becomes established and normalized, the pressure to use it for additional purposes will grow and that the oversight mechanisms are not robust enough to contain that expansion.

This concern reflects a broader pattern in surveillance law: authorities established for specific purposes tend to expand as the infrastructure they create becomes available and as new applications are identified. The PATRIOT Act's telephone metadata collection program, which was legally justified under a provision designed for limited pen register surveillance, is the most prominent recent example. Critics of CISA argued that its architecture was similarly susceptible to authority creep — that what begins as cybersecurity threat sharing could evolve into a broad general-purpose surveillance capability with financial sector data at its core.

The table below summarizes the key dimensions of the CISA debate in a format useful for debate preparation.

Dimension	The Case For CISA	The Case Against CISA
Core rationale	Cyber threats evolve faster than any single firm can track alone; collective defense requires shared intelligence; the private sector sees the attacks, the government has the context	Voluntary information sharing already existed; CISA's primary effect is expanding government access to private sector data under a security label
Privacy protections	CISA requires scrubbing of personal information before sharing; DHS must review	Scrubbing is imperfect and automated; threat indicators often contain IP

	shared data and remove personally identifiable information before passing it to other agencies	addresses, behavioral patterns, and content that can identify individuals; no independent verification that scrubbing is adequate
Government surveillance	Sharing is voluntary; firms choose what to share; government cannot compel sharing under CISA	Data shared for cybersecurity purposes can be used for other government functions under broad exceptions; CISA effectively creates a new surveillance data stream with limited oversight
Liability protections	Without liability protection, firms faced legal risk for sharing threat data; the liability shield was the essential mechanism enabling participation	Liability protections are too broad; they shield firms from accountability for sharing excessive personal data and potentially from private lawsuits unrelated to the sharing itself
Effectiveness	Financial sector threat sharing through FS-ISAC demonstrates measurable benefit; real-time automated sharing via AIS has increased the speed of collective defense; nations that share threat intelligence collectively have faster response times	Most shared indicators are already known or stale by the time they are shared; sophisticated nation-state actors change tactics faster than shared indicators can be acted upon; the benefit is concentrated in large firms with resources to process shared data
Oversight	DHS oversight, privacy guidelines, and annual reporting to Congress provide accountability; Inspector General reviews	Oversight mechanisms are internal to the executive branch; no independent judicial review of how shared data is used; Congress receives reports but has limited operational visibility
Civil liberties balance	The law targets threat indicators, not	The distinction between threat indicators and

	communications content; it does not authorize surveillance of American citizens	content is blurry in practice; the law's exceptions are broad enough to encompass significant personal data; the civil liberties community broadly opposed the law for these reasons

The Broader Legal Framework for Cybersecurity

CISA addresses one dimension of cybersecurity law — information sharing. The full legal framework governing cybersecurity obligations for FinTech firms is considerably broader. Understanding it requires knowing the key statutes, regulatory requirements, and voluntary standards that together define what FinTech firms are legally required to do, what they are encouraged to do, and what the consequences of failure look like.

The Computer Fraud and Abuse Act

The Computer Fraud and Abuse Act, or CFAA, enacted in 1986 and amended multiple times since, is the primary federal statute criminalizing unauthorized access to computer systems. The CFAA makes it a federal crime to intentionally access a computer without authorization or to exceed authorized access and obtain information — a broad prohibition that covers most forms of hacking, computer intrusion, and unauthorized data access. Violations carry penalties ranging from fines to imprisonment, with enhanced penalties for attacks on critical infrastructure or attacks that cause significant damage.

For FinTech firms, the CFAA is relevant in two directions. As a victim of cybercrime, a FinTech firm can bring CFAA civil claims against attackers — a mechanism that has been used by companies to pursue hackers, former employees who misused system access, and competitors who accessed systems improperly. As an operator of information systems, a FinTech firm must ensure that its own practices — including security testing, bug bounty programs, and access control implementations — do not inadvertently create CFAA exposure for its own employees or authorized researchers.

The CFAA's prohibition on exceeding authorized access has been the subject of significant legal debate, because its scope determines whether employees who misuse data they are authorized to access — a significant category of insider threat — are committing federal crimes or merely civil wrongs. The Supreme Court's 2021 decision in Van Buren v. United States narrowed the CFAA's scope, holding that exceeding authorized access means accessing areas of a computer system to which one's existing authorization does not extend, not using legitimately accessible information for improper purposes. This narrowing has implications for how insider threat cases are prosecuted and for the legal tools available to FinTech firms responding to insider data theft.

Sector-Specific Regulatory Requirements

FinTech firms face cybersecurity obligations not just from general cybersecurity law but from the sector-specific regulatory frameworks that govern their particular activities. The GLBA Safeguards Rule, covered in Chapter 8, requires financial institutions to implement written information security programs and has been substantially updated to include specific technical requirements including encryption, multi-factor authentication, and breach notification. FinTech firms that qualify as financial institutions under GLBA must comply with these requirements as a condition of their regulatory status.

The federal banking regulators — the OCC, the Federal Reserve, and the FDIC — have issued interagency cybersecurity guidance and examination procedures that apply to supervised financial institutions and their service providers. These regulators examine the cybersecurity programs of the banks and financial companies they supervise, assess compliance with regulatory expectations, and can take enforcement action against institutions with inadequate cybersecurity controls. For FinTech firms that partner with banks or that are directly supervised by banking regulators, these examination standards are a significant compliance driver.

State regulators have also developed cybersecurity requirements for financial firms. The New York Department of Financial Services, or NYDFS, enacted its Cybersecurity Regulation in 2017 — one of the first comprehensive state cybersecurity rules applicable to financial services firms. The NYDFS regulation requires covered entities to maintain a cybersecurity program, appoint a Chief Information Security Officer, implement specific technical controls, conduct annual penetration testing, and report cybersecurity incidents to the regulator within 72 hours. The NYDFS regulation has been updated multiple times since its initial adoption and has served as a model for other states developing cybersecurity requirements for financial firms.

The NIST Cybersecurity Framework

The National Institute of Standards and Technology Cybersecurity Framework, developed in response to a 2013 executive order and updated most recently in 2024 as version 2.0, is a voluntary framework that provides a structured approach to managing cybersecurity risk. The framework is organized around six core functions: Govern, Identify, Protect, Detect, Respond, and Recover. Each function encompasses categories and subcategories of cybersecurity activities, providing a common language and structure for discussing, assessing, and improving cybersecurity programs.

The NIST framework is voluntary — no federal law requires private sector firms to adopt it. But it has become the de facto standard for cybersecurity program design in the United States, referenced by federal regulators, insurance underwriters, and contract counterparties as the benchmark against which security programs are evaluated. A FinTech firm that can demonstrate its security program is aligned with the NIST framework has a credible basis for asserting that it meets reasonable cybersecurity standards, which is relevant to regulatory examinations, litigation defense, and cyber insurance pricing.

The framework's Respond and Recover functions are particularly relevant to the incident response discussion in Chapter 9 — they provide a structured approach to building the capabilities that a firm needs when an attack occurs. The Govern function, added in version 2.0, reflects the growing recognition that cybersecurity is a governance and organizational culture issue as much as a technical one — a message consistent with the security culture discussion at the end of Chapter 9.

The Roles of Government and the Private Sector

One of the most contested dimensions of cybersecurity policy is the appropriate division of responsibility between government and the private sector. Understanding this debate is important for anyone engaged with cybersecurity policy in a professional capacity.

The Private Sector's Primary Role

The vast majority of U.S. critical infrastructure — including financial market infrastructure, payment networks, and the technology platforms on which FinTech services run — is owned and operated by private companies. This means that the first line of defense against most cybersecurity threats is the private sector, because the systems being attacked belong to private firms that have the primary responsibility and authority to protect them. Government cannot unilaterally install security controls in privately owned systems or require specific security architectures without legislation authorizing it to do so.

This structural reality means that the private sector's cybersecurity investment, practices, and culture determine the baseline security level of the systems that adversaries target. A government cybersecurity program, however well-resourced and capable, cannot compensate for a private sector that does not take security seriously. The security investment decisions made by FinTech firms — about staffing, architecture, tooling, and vendor selection — have direct implications for the security of the financial infrastructure on which millions of people depend.

The Government's Unique Capabilities

Government brings capabilities to cybersecurity that the private sector does not have. Intelligence agencies collect signals intelligence about adversary intentions, capabilities, and operations that is not available through commercial means. Law enforcement agencies can investigate cybercrimes, execute search warrants, seize criminal infrastructure, extradite suspects, and prosecute attackers. Diplomatic channels can be used to pressure foreign governments to take action against cybercriminals operating within their jurisdictions or to negotiate norms of behavior in cyberspace. And in the most extreme cases, military capabilities can be brought to bear in response to cyber attacks.

The Department of Defense has stated explicitly that it reserves the right to respond to significant cyberattacks with any means necessary, including military force. This principle — that the government may respond with military force as a reaction to a cyberattack — establishes that cyber operations are not categorically different from other forms of attack in terms of the responses they may justify. A cyberattack that causes the kind of destruction or disruption that a conventional military attack would cause can legitimately be treated as an act of war, with all that implies about the available responses.

CISA — the agency, the Cybersecurity and Infrastructure Security Agency, which was created in 2018 partly in response to the legislative

CISA framework — serves as the civilian coordinator of the government's cybersecurity activities. It provides technical assistance, threat intelligence, and incident response support to private sector firms, particularly those operating critical infrastructure. Its advisories, alerts, and resources are available to FinTech firms as part of the government's contribution to collective defense.

The Information Asymmetry Problem

The most significant structural challenge in the government-private sector relationship is information asymmetry running in both directions. The private sector sees the attacks — financial institutions experience millions of intrusion attempts daily and develop detailed operational knowledge of the techniques being used against them. The government has the context — intelligence about who the attackers are, what they are trying to accomplish, and what vulnerabilities they are targeting based on classified sources that the private sector cannot access.

Effective national cybersecurity requires bridging this asymmetry. CISA's information sharing framework is one mechanism for doing so — it creates a legal and operational infrastructure for the private sector's attack knowledge to flow to the government and for the government's threat context to flow back. But the asymmetry is only partially addressed by information sharing. The government's classified intelligence is not fully shareable even through a framework like CISA — much of it is classified at levels that prevent broad distribution. And the private sector's operational knowledge is fragmented across thousands of individual firms that do not share with each other as freely as CISA's proponents hoped.

The NSA's Dual Role

One of the persistent tensions in U.S. cybersecurity policy is the role of the National Security Agency, which has both offensive and defensive intelligence missions. The NSA's offensive mission involves developing capabilities to penetrate adversary computer systems — capabilities that require discovering and maintaining knowledge of software vulnerabilities. The NSA's defensive mission involves protecting U.S. government and critical infrastructure systems — which would be served by disclosing those same vulnerabilities to software vendors so they can be patched.

These missions are in direct tension. A vulnerability that the NSA has developed an offensive capability around cannot simultaneously be disclosed for patching without eliminating the offensive capability. The NSA's process for deciding which vulnerabilities to disclose and which to retain for offensive use — the Vulnerabilities Equities Process — has been criticized by security researchers and civil liberties advocates who argue that the decision systematically favors offensive capability over defensive patching. The 2017 WannaCry ransomware attack, which used a vulnerability that the NSA had identified and retained, and which affected hundreds of thousands of systems worldwide after the vulnerability was leaked, illustrated the real-world cost of retaining rather than disclosing vulnerabilities.

This tension is relevant to the CISA debate because it illustrates a broader point that CISA's critics make: the government's cybersecurity interests are not perfectly aligned with the private sector's. The government's intelligence mission sometimes leads it to retain vulnerabilities, withhold threat information, and prioritize offensive capabilities in ways that reduce rather than enhance private sector security. Designing information sharing frameworks that genuinely serve both sets of interests requires grappling with this misalignment honestly.

Balancing Security and Civil Liberties

The tension between cybersecurity and civil liberties is not unique to CISA — it runs through every cybersecurity policy debate. Understanding how to think about this tension is important for professional engagement with cybersecurity policy questions throughout a career in FinTech.

The Nature of the Tension

Effective cybersecurity often requires surveillance — monitoring network traffic, analyzing behavioral patterns, tracking the movement of threat indicators across systems, and correlating events across multiple sources to identify intrusions. Many of these activities, when conducted at scale and on data that flows through civilian infrastructure, collect information about innocent people who are not attackers. The surveillance capability that detects a sophisticated intrusion attempt may be indistinguishable, at the technical level, from the surveillance capability that monitors the communications and financial activity of ordinary citizens.

This is not a hypothetical concern. The NSA's bulk metadata collection program, the FBI's use of National Security Letters to access financial records without judicial approval, and the broad subpoena authority that government agencies use to access data held by FinTech firms all represent points at which government cybersecurity and intelligence capabilities intersect with the private financial information of ordinary people. Each of these programs has defenders who argue it is necessary for national security and critics who argue it violates constitutional rights.

The Fourth Amendment in the Cybersecurity Context

The Fourth Amendment's protection against unreasonable searches and seizures applies to government action — it constrains what the government can do with information it collects through surveillance, not what private companies can collect on their own. But as Chapter 8 discussed in connection with Carpenter v. United States, the scope of Fourth Amendment protection in the digital context is actively evolving. The principle that the comprehensiveness of a surveillance capability matters to its constitutional status — that a dataset comprehensive enough to reveal the privacies of life deserves stronger constitutional protection than a limited record — has direct implications for cybersecurity surveillance programs that aggregate data across many sources.

The specific civil liberties concern about CISA is that shared cyber threat indicators, which may include IP addresses, behavioral patterns, and network metadata, can be used to build profiles of individuals even after the scrubbing process. If those profiles are maintained in government databases and associated with individuals' other records, the result may be exactly the kind of comprehensive digital surveillance that Carpenter's logic would suggest warrants constitutional scrutiny. CISA's use limitations are designed to prevent this outcome, but critics argue those limitations are insufficiently robust to provide meaningful protection.

How CISA Tries to Balance the Tension

CISA's architects made several specific choices designed to address the security-civil liberties tension. The civilian DHS architecture rather than NSA or military routing was a deliberate choice to keep the program outside the intelligence community's oversight framework and within civilian accountability structures. The scrubbing requirement was designed to limit personal information flows. The use restrictions were designed to prevent security-labeled data from becoming a general-purpose surveillance tool. And the oversight mechanisms — Privacy and Civil Liberties Oversight Board consultation, Inspector General review, congressional reporting — were designed to create external accountability.

Whether these mechanisms are adequate is the core question on which the CISA debate turns. Proponents argue they represent a reasonable balance — a framework that enables meaningful security information sharing while maintaining meaningful privacy protections and accountability. Critics argue they are insufficient — that the scrubbing is imperfect, the use restrictions are too broad, the oversight is too internal, and the potential for abuse is too significant to accept on the basis of the security benefits the law is likely to produce.

This disagreement is genuine and unresolved. It is a disagreement about facts — how much personal information actually flows through CISA despite the scrubbing requirement, how effectively the use limitations constrain government behavior, and how much the sharing actually improves collective cybersecurity. It is also a disagreement about values — how much weight to assign to security versus privacy, and who bears the burden of proof when those values conflict. Anyone engaging seriously with CISA will encounter this tension repeatedly, and the quality of that engagement depends in part on how honestly and carefully both sides of it are considered.

Key Takeaways

The Cybersecurity Information Sharing Act of 2015 was enacted to address a structural asymmetry in cybersecurity: attackers share tools and techniques freely while defenders operated in isolation. CISA creates a voluntary framework for sharing cyber threat indicators and defensive measures between private entities and between the private sector and the federal government, with liability protections for participating firms and privacy safeguards designed to limit personal information flows.

CISA's core provisions include: authorization for private entities to monitor their own information systems for cybersecurity purposes; authorization to share cyber threat indicators and defensive measures with other private entities and with DHS; a requirement to scrub personal information before sharing; real-time automated sharing through the DHS Automated Indicator Sharing platform; liability protection for compliant sharing; use restrictions limiting government use of shared data to cybersecurity and specified law enforcement and national security purposes; and oversight mechanisms including Privacy and Civil Liberties Oversight Board consultation, Inspector General review, and congressional reporting.

The case for CISA rests on: the collective defense argument — that individual firm defense is insufficient against threats that target the entire sector simultaneously; the liability shield as the mechanism that enables sharing by removing the legal risk that previously deterred it; the leveling effect that makes sector-wide threat intelligence available to smaller firms that cannot develop it independently; and the government-to-private sector intelligence channel that makes classified threat context available to private defenders.

The case against CISA rests on: the privacy objection that scrubbing is insufficient and that shared indicators contain personal information that builds government surveillance profiles; the surveillance expansion argument that CISA functions as a broad government data collection program under a security label; the effectiveness question that most shared indicators are low-value and that the most dangerous adversaries are not countered by indicator sharing; and the scope creep concern that CISA's broad authorities will expand over time as the information sharing infrastructure becomes established.

The broader legal framework governing cybersecurity for FinTech firms includes the Computer Fraud and Abuse Act as the primary federal cybercrime statute; sector-specific obligations under GLBA, banking regulator guidance, and the NYDFS Cybersecurity Regulation; and the NIST Cybersecurity Framework as the de facto voluntary standard for security program design. Together these create a layered compliance environment that FinTech firms must navigate alongside the voluntary CISA participation framework.

The appropriate roles of government and the private sector in national cybersecurity reflect a structural reality: the private sector owns and operates most of the critical infrastructure being attacked and bears primary responsibility for its defense, while the government brings unique capabilities including intelligence, law enforcement, diplomacy, and military force. The government may respond to significant cyberattacks with military force — a principle that establishes cyber operations as within the spectrum of acts that can justify kinetic military responses. CISA — both the law and the agency — represents the primary institutional mechanism for connecting private sector operational knowledge with government capabilities in pursuit of collective defense.

The tension between cybersecurity and civil liberties runs through all cybersecurity policy. Effective threat detection requires surveillance capabilities that also collect data about innocent people. CISA attempts to balance this tension through scrubbing requirements, use restrictions, and oversight mechanisms. Whether those mechanisms are adequate is a genuine policy disagreement that turns on both empirical questions about how the law operates in practice and normative questions about the appropriate weight of security versus privacy when they conflict.

Chapter 11: Risk Management and Internal Controls in FinTech

Introduction

Every financial services firm operates under conditions of uncertainty. Markets move unexpectedly. Systems fail. People make mistakes or act in bad faith. Regulations change. Competitors innovate. Customers default. Attackers probe for weaknesses. The question for a FinTech firm is not whether these things will happen but how well-prepared it is when they do.

Risk management is the discipline of identifying, assessing, and responding to those uncertainties in a structured way. It is not the elimination of risk — that is neither possible nor desirable, because taking risk is how financial firms generate returns. It is the deliberate management of risk: understanding what could go wrong, evaluating how likely and how severe each possibility is, and putting in place the controls that reduce the probability and impact of adverse outcomes to acceptable levels.

Internal controls are the mechanisms through which risk management is implemented. They are the policies, procedures, systems, and practices that an organization uses to prevent problems from occurring, detect them when they do, and correct them before they cause lasting damage. For FinTech firms, whose operations are entirely digital, whose customer relationships depend on trust, and whose regulatory environments carry significant compliance obligations, internal controls are not optional infrastructure — they are fundamental to the business.

This chapter covers the risk management framework from the ground up. It begins with why FinTech firms face a higher overall risk profile than traditional financial firms. It maps the major types of risk facing FinTech companies. It explains the qualitative risk assessment process that practitioners use to identify, evaluate, and prioritize risks in an organizational context. And it covers the three types of internal controls — preventative, detective, and corrective — in the depth required to design them, classify them correctly, and explain their trade-offs. Chapter 12 builds on this foundation to cover corporate governance, anti-money laundering, know-your-customer requirements, and the governance failures that have enabled some of the most significant FinTech collapses.

Why Risk Is Higher in FinTech

Overall risk is generally higher in FinTech firms than in traditional financial firms. This is not a criticism of FinTech — it reflects the structural characteristics of the sector. Understanding why the risk profile is elevated is essential for designing risk management programs that are actually calibrated to the environment FinTech firms operate in.

Traditional banks and financial institutions have built their risk management capabilities over decades, in some cases over centuries. They have mature compliance functions, established control frameworks, experienced audit teams, and regulatory examination schedules that impose external discipline on their risk management practices. They operate under capital adequacy requirements that force them to hold reserves against unexpected losses. They have physical infrastructure — branch networks, vaults, paper records — that provides redundancy and fallback when digital systems fail.

FinTech firms typically have none of this institutional heritage. They are built for speed and scale, not for the methodical accumulation of risk management capability. Their founding teams are dominated by technologists and product designers, not compliance officers and risk managers. Their cultures prioritize shipping product and acquiring customers, and the controls that might slow down those objectives are viewed as costs to be minimized. The result is a risk profile that is elevated across virtually every category — market, credit, operational, and business risk are all generally higher in FinTech firms than in their traditional counterparts.

FinTech startups, in particular, tend to worry less about security than traditional banks. This is a documented pattern, not an unfair generalization. A startup in its first two years is spending its capital on product development, marketing, and growth. Its team is small, stretched, and focused on proving the business model. The compliance and security investments that a large bank makes as a matter of course — dedicated compliance staff, regular external audits, formalized information security programs, board-level risk committees — are absent. The startup may have no dedicated security professional at all, no formal risk assessment process, and controls that were designed to be revisited when the firm reaches scale but that in practice never get the attention they need.

The risk elevation in FinTech is also structural. Digital-first businesses have no analog fallback — if the app is down, the service is down, with no physical branch to absorb customer demand. They rely on complex ecosystems of third-party vendors, cloud providers, and API partners, each of which introduces dependencies and potential failure points. They collect sensitive financial data at scale and often before they have built adequate infrastructure to protect it. And they operate in regulatory environments that are evolving quickly, creating compliance risk that established firms, with their lobbying capacity and regulatory relationships, are better positioned to navigate.

Types of Risk Facing FinTech Firms

Risk in FinTech does not come from a single direction. It comes from many directions simultaneously, and an effective risk management program requires understanding each category well enough to identify specific risks within it. The table below provides an overview of the major risk types facing FinTech firms, with definitions, examples, and the FinTech-specific factors that elevate each.

Risk Type	Definition	FinTech Examples	Why Higher in FinTech
Operational Risk	Risk of loss from inadequate or failed internal processes, people, systems, or external events	System outages during peak trading; algorithm errors executing wrong trades; employee fraud; third-party vendor failure; cyberattack disrupting payment processing	Digital-first architecture creates single points of failure; rapid scaling outpaces process maturity; heavy reliance on third-party infrastructure; complex API ecosystems multiply failure points
Market Risk	Risk of loss from changes in market prices, interest rates, exchange rates, or asset values	Cryptocurrency price volatility affecting exchange reserves; interest rate changes affecting lending platform margins; foreign exchange exposure for cross-border payment platforms	Many FinTech firms hold crypto or digital assets whose volatility exceeds traditional financial instruments; algorithmic trading amplifies market moves; limited hedging infrastructure compared to established banks
Credit Risk	Risk that a borrower or counterparty will fail to meet its obligations	Borrower defaults on peer-to-peer lending platform loans; buy-now-pay-later customers failing to repay; counterparty	FinTech lenders often serve underbanked populations with limited credit history; alternative credit scoring models are less tested; BNPL

		failure in crypto settlement	platforms may extend credit with minimal underwriting
Business Risk	Risk of loss from adverse effects of operating in different markets, regulatory environments, or competitive conditions	New regulation restricting a core business model; competitor innovation making a product obsolete; loss of a key banking partner; adverse regulatory ruling in a target market	FinTech firms often operate in regulatory gray areas subject to sudden clarification; business models may depend on single regulatory interpretations; international expansion exposes firms to unfamiliar regulatory regimes
IT / Technology Risk	Risk of loss from technology failures, inadequate systems, or technological obsolescence	Core banking system outage; data corruption; software bugs causing incorrect calculations; legacy code creating security vulnerabilities	Technology is the entire delivery mechanism; no physical branch network as fallback; continuous deployment cycles introduce new vulnerabilities; technical debt from rapid growth creates hidden risks
Compliance Risk	Risk of legal penalties, financial loss, or reputational damage from failure to comply with applicable laws and regulations	AML/KYC failures; unlicensed money transmission; inadequate consumer disclosures; violation of state-specific licensing requirements	Regulatory landscape is fragmented and evolving; FinTech often operates across multiple jurisdictions simultaneously; compliance functions frequently

			understaffed relative to business growth
Reputational Risk	Risk of damage to the firm's standing with customers, partners, regulators, or the public	High-profile data breach; visible customer service failures; association with fraud or illegal activity; regulatory enforcement action	Digital-first firms are highly visible; social media amplifies incidents rapidly; customer trust is the primary asset and is hard to rebuild once lost

Operational Risk

Operational risk deserves particular attention because it is the category most distinctive to FinTech's operating model. Traditional definitions of operational risk — loss from inadequate or failed internal processes, people, systems, or external events — were developed primarily in the context of large banks with complex manual processes. In a FinTech context, operational risk is dominated by technology failure, third-party dependency, and the process gaps that result from rapid growth without commensurate investment in operational maturity.

The single most common source of operational risk in FinTech is the failure of systems that customers depend on for access to their money or financial services. A payment app that goes down during peak usage hours — Black Friday, the end of a pay cycle, a market event that drives high trading volume — exposes customers to real harm and the firm to regulatory scrutiny, customer attrition, and reputational damage. Unlike a bank branch that can handle overflow manually, a digital platform has no offline mode. When the system is down, service is down.

Operational risk also encompasses the risks created by relying on third-party infrastructure. A FinTech firm whose payment processing depends on a single provider, whose identity verification relies on one vendor, and whose cloud infrastructure runs on one platform has concentrated its operational risk in ways that are often not visible until a vendor experiences a significant outage. The 2021 Fastly content delivery network outage, which took down large portions of the internet for approximately an hour, illustrated how deeply embedded shared infrastructure has become in digital businesses and how concentrated the operational risk of that dependence can be.

Market Risk in FinTech

Market risk for most traditional financial firms centers on interest rate movements, equity price changes, and foreign exchange fluctuations — risks that are well-understood and managed through established hedging instruments. For FinTech firms, market risk takes those traditional forms but is amplified by the sector's heavy involvement with cryptocurrency and the algorithmic trading and automated investment products that FinTech firms offer.

Cryptocurrency's extreme price volatility creates market risk that is qualitatively different from equity or bond market movements. A cryptocurrency exchange that holds a reserve of digital assets to facilitate customer transactions is exposed to rapid value changes in those assets. A crypto lending platform whose collateral consists of digital assets can find that collateral has lost a substantial fraction of its value before it can act to protect itself. The 2022 collapse of the Terra/Luna stablecoin ecosystem wiped out tens of billions of dollars in value within days — a speed of deterioration with no parallel in traditional financial markets.

Credit Risk in the FinTech Context

FinTech firms that extend credit — through peer-to-peer lending platforms, buy-now-pay-later products, merchant cash advances, or direct consumer loans — face credit risk: the possibility that borrowers will not repay. The distinctive credit risk challenge for FinTech lenders is that many of them were explicitly built to serve customers that traditional lenders had excluded — people with thin credit files, gig economy workers with irregular income, small businesses without the collateral that bank lenders require.

Serving underbanked populations is both a genuine social good and a significant source of credit risk. The alternative credit scoring models that FinTech lenders use to assess these borrowers — incorporating transaction data, behavioral signals, and non-traditional data sources — may be less tested through a full economic cycle than the credit bureau scores that traditional lenders rely on. When the economic environment deteriorates, models trained primarily on good-times data may underestimate default rates significantly.

The buy-now-pay-later sector illustrates this risk in practice. BNPL platforms extended credit to millions of consumers through a checkout experience that minimized friction and required minimal underwriting. As interest rates rose and economic conditions tightened in 2022 and 2023, delinquency rates at several major BNPL providers increased significantly — raising questions about whether the ease of access that made the product attractive had also extended credit to consumers who would not repay.

Business Risk

Business risk — the risk of loss from adverse changes in the environment in which a firm operates — is elevated in FinTech because so many FinTech business models depend on specific regulatory interpretations, competitive dynamics, or distribution relationships that can change rapidly. A FinTech firm whose entire business model depends on a charter-lite approach to banking services — offering deposit-like products without a banking license, relying on a bank partner for regulatory coverage — faces existential business risk if that partnership is terminated, if the bank partner fails, or if regulators clarify that the arrangement is impermissible.

The Synapse Financial Technologies collapse in 2024 is a recent illustration of this risk. Synapse was a banking-as-a-service middleware provider that connected FinTech apps to their bank partners, handling the technical layer between consumer-facing FinTech platforms and the insured deposits held at partner banks. When Synapse filed for bankruptcy in May 2024, its collapse revealed that customer funds held through multiple FinTech platforms could not be fully reconciled — the accounting records between Synapse, its FinTech clients, and the underlying banks did not match, leaving customers unable to access their money for months. The business risk that materialized was not simply Synapse's own failure but the cascading effects of that failure on every FinTech firm that had built its product on Synapse's infrastructure.

The Risk Management Process

Identifying that risks exist is the beginning of risk management, not the end. A structured risk management process moves through several interconnected stages: identifying risks, assessing them, prioritizing responses, implementing controls, and monitoring effectiveness over time. This process provides the practical framework for conducting a qualitative risk assessment in any organizational context.

Risk Identification

Risk identification is the systematic process of cataloguing everything that could go wrong. Effective risk identification is comprehensive rather than selective — it covers technical risks and non-technical ones, internal threats and external ones, low-probability catastrophic risks and high-probability nuisance risks. A thorough risk inventory for a FinTech firm should include both technical security risks like phishing and system outages, and non-technical ones like regulatory change, key person dependency, and vendor failure.

A useful organizing framework for risk identification is threat origin — where does the threat come from? Threats originate from several directions: internal sources including employees, contractors, and authorized system users; external sources including hackers, criminals, competitors, and regulators; natural causes including weather events, earthquakes, and pandemics that disrupt operations; accidental causes including human error, software bugs, and unintentional misconfiguration; and malicious causes including deliberate fraud, sabotage, and targeted attack. Organizing risks by threat origin helps ensure that identification covers the full range of potential sources rather than focusing exclusively on the most visible category.

For a FinTech firm conducting a risk identification exercise, the result should be a list that spans multiple categories and threat origins. A payment platform might identify: phishing attacks targeting employees (external, malicious); misconfiguration of cloud infrastructure (internal, accidental); a key vendor's financial failure (external, non-malicious); interest rate changes reducing the value of the firm's lending portfolio (market risk, external); an employee exfiltrating customer data before departure (internal, malicious); a regulatory ruling requiring changes to the firm's product (external, regulatory); and a prolonged system outage following a major software update (operational, accidental). Each of these is a distinct risk requiring its own assessment and response.

Risk Assessment: Likelihood and Impact

Once risks have been identified, each must be assessed on two dimensions: likelihood and impact. Likelihood is the probability that the risk event will actually occur within a given time period. Impact is the severity of the consequences if the risk event does occur — its effect on the firm's finances, operations, customers, and reputation. The product of likelihood and impact is the risk level, which determines how urgently the risk requires a response.

Qualitative risk assessment uses descriptive scales rather than precise numerical probabilities. A common approach assigns likelihood ratings of low, medium, or high based on the frequency with which the risk is likely to materialize and impact ratings on a similar scale based on the severity of the consequences. The resulting risk matrix — low likelihood

/ low impact through high likelihood / high impact — produces a prioritization framework that allows a firm to direct its control investment toward the risks that pose the greatest combined threat.

A phishing attack targeting a FinTech firm's employees might be rated high likelihood (phishing attacks are frequent and improving in sophistication) and medium impact (a successful attack on one employee's credentials is damaging but not catastrophic if proper controls limit the lateral movement an attacker can achieve). A major earthquake destroying the firm's primary data center might be rated low likelihood and high impact. A routine software bug causing minor transaction errors might be rated medium likelihood and low impact. Each assessment drives a different control response.

Scoring each identified risk on likelihood, impact, and the resulting risk level is the mechanism that turns a risk inventory into a prioritization tool. The purpose is not to achieve actuarial precision — qualitative assessment is explicitly an estimate, not a calculation — but to develop the habit of thinking about risks in terms of their comparative priority rather than treating all risks as equally urgent or equally ignorable.

Risk Response: Mitigation, Transfer, Acceptance, and Avoidance

Having assessed a risk, a firm has four basic response options. Risk mitigation involves implementing controls to reduce either the likelihood that the risk event occurs or the impact if it does — this is the most common response and the focus of the internal controls discussion below. Risk transfer involves shifting the financial consequences of the risk to a third party, most commonly through insurance — a firm that purchases cyber insurance is transferring some of the financial impact of a data breach to the insurer. Risk acceptance means acknowledging that the risk exists and deciding not to take specific action to reduce it, typically because the cost of mitigation exceeds the expected value of the harm — a deliberately chosen position, not an oversight. Risk avoidance means declining to engage in an activity that creates the risk — a FinTech firm that decides not to offer cryptocurrency trading services because the regulatory and market risks are unacceptable is avoiding those risks.

Most real-world risk management involves a combination of these responses. A FinTech firm might mitigate cybersecurity risk through technical controls and security training, transfer residual risk through cyber insurance, accept the risk of minor operational disruptions whose mitigation cost exceeds their expected impact, and avoid certain high-risk product categories entirely. The goal is not to eliminate risk but to manage it to a level that is acceptable given the firm's risk appetite — the amount of risk it is willing to take in pursuit of its business objectives.

Internal Controls: The Three Types

Internal controls are the specific mechanisms through which risk responses are implemented. They are not abstractions — they are concrete policies, procedures, systems, and practices that exist in the real operations of a FinTech firm. Understanding the three types of internal controls — preventative, detective, and corrective — and being able to design, classify, and evaluate them is a core competency for anyone working in FinTech risk management.

The classification matters because the three types serve different purposes, operate at different times relative to the risk event, and have different cost-benefit profiles. A strong control program includes all three types working together: preventing what can be prevented, detecting what cannot be prevented, and correcting what cannot be detected in time to prevent. The table below provides an overview before each type is discussed in depth.

Control Type	Purpose	Timing	FinTech Examples
Preventative	Stop a risk event from occurring in the first place	Before the event	Multi-factor authentication; role-based access controls and least privilege; input validation on transaction

			forms; segregation of duties in fund transfer approval; employee background checks; mandatory security awareness training; encryption of data at rest and in transit; transaction velocity limits
Detective	Identify that a risk event has occurred or is occurring	During or immediately after the event	Real-time transaction monitoring for anomalous patterns; audit logs recording all system access and changes; automated reconciliation of account balances; intrusion detection and SIEM alerts; exception reports flagging unusual activity; internal audit reviews; penetration testing; whistleblower hotlines
Corrective	Fix the problem after it has been identified; restore normal operations; prevent recurrence	After the event	Incident response and business continuity plans; data backup and disaster recovery systems; patch management for discovered vulnerabilities; customer notification and remediation procedures; disciplinary action for policy violations; root cause analysis and process redesign; cyber insurance; regulatory breach reporting procedures

Preventative Controls

A preventative control is designed to stop a risk event from occurring in the first place. It acts before the event, reducing the likelihood that the risk materializes. Preventative controls are generally the most desirable type because they are the most efficient — it is cheaper and less damaging to prevent a problem than to detect and correct it. But prevention is not always possible, and preventative controls always involve trade-offs between security and usability.

Multi-factor authentication is a preventative control. By requiring a second verification factor beyond a password, it reduces the likelihood that stolen credentials result in unauthorized account access. It does not eliminate the risk entirely — SIM-swapping attacks can defeat SMS-based MFA, and social engineering can sometimes circumvent even strong authentication — but it substantially raises the cost and difficulty of unauthorized access. The trade-off is the friction it imposes on legitimate users and the support burden created by users who lose access to their second factor.

Role-based access controls and the principle of least privilege are preventative controls that limit what any given user or system component can do. By restricting each employee's access to only the systems and data required for their specific role, least privilege reduces the likelihood that a compromised account or a malicious insider can access sensitive data or take damaging actions. An employee in the customer service function who can view account balances but cannot initiate transfers cannot cause the same damage as one with unrestricted transaction access. The trade-off is the operational complexity of managing many different access levels and the friction that results when employees need occasional access to systems outside their normal scope.

Segregation of duties is one of the most important preventative controls in financial services. It requires that critical transactions or functions be divided among multiple people so that no single individual can complete a high-risk action alone. In a payment platform, segregation of duties might require that initiating a wire transfer and approving it be performed by different employees — so that a fraudulent transfer cannot be completed by one person acting alone. In an accounting function, it might require that the person who records transactions be different from the person who reconciles accounts. The control prevents both intentional fraud and accidental error by requiring multiple parties to be complicit in or to review the same action.

Employee background checks conducted before hiring, mandatory security awareness training, and documented access provisioning and de-provisioning procedures are all preventative controls focused on the human dimension of risk. They reduce the likelihood that employees will introduce risk through negligence, poor judgment, or deliberate misconduct. Transaction velocity limits — rules that flag or block transactions above certain thresholds — are preventative controls that reduce the likelihood that automated fraud or errors result in large financial losses before they are detected.

Input validation on transaction forms and APIs is a technical preventative control that reduces the likelihood of injection attacks, data corruption, and processing errors by verifying that data entered into a system conforms to expected formats and ranges before it is processed. A payment form that validates that an account number is the correct length and format before submitting it prevents a category of both fraud and error. Encryption of data at rest and in transit is a preventative control that reduces the likelihood that intercepted or stolen data can be read and used by an attacker.

Detective Controls

A detective control identifies that a risk event has occurred or is in progress. Where preventative controls reduce the likelihood of the event occurring, detective controls reduce the impact by enabling faster response. The faster a problem is detected, the less damage it causes — a fraudulent transaction detected and reversed within minutes causes less harm than one that goes unnoticed for weeks.

Real-time transaction monitoring is one of the most valuable detective controls in FinTech. Automated systems that analyze every transaction against behavioral baselines and pattern rules can flag anomalies for immediate review — a transfer of an unusual amount to an unfamiliar account, a login from an unfamiliar location, a sequence of small transactions that together constitute unusual activity. These systems generate alerts that route to fraud teams or trigger automated responses such as temporarily blocking the transaction pending customer verification. The effectiveness of transaction monitoring depends on the quality of the underlying rules and models — poorly calibrated systems generate high false-positive rates that overwhelm fraud teams and erode customer experience.

Audit logs — records of all system access, configuration changes, and transactions — are a foundational detective control. They create an evidence trail that can be reviewed after an incident to understand what happened, who did it, and how the attack or error progressed through the system. Audit logs are only useful as detective controls if they are reviewed — an audit log that is generated but never examined is not providing any detection value. Automated log analysis, including security information and event management systems that correlate events across multiple sources, turns audit logs from a passive record into an active detection capability.

Reconciliation is a detective control with deep roots in traditional accounting practice. Automated reconciliation of account balances — comparing what the system shows against what the underlying records indicate, on a scheduled basis — detects discrepancies that may indicate fraud, error, or system malfunction. A payment platform that reconciles its processing records against bank settlement data daily will detect discrepancies far faster than one that reconciles monthly, and will be able to investigate and correct them before they compound.

Penetration testing — engaging security professionals to attempt to compromise a firm's systems using the same techniques that real attackers would use — is a detective control that identifies vulnerabilities before attackers do. A penetration test that discovers an API endpoint returning more customer data than it should reveals a vulnerability that can be fixed before it is exploited. The limitation of penetration testing as a detective control is its periodic nature — a test conducted once a year does not detect vulnerabilities introduced in the months since the last test. Continuous vulnerability scanning provides more timely detection but at the cost of significant noise from low-severity findings.

Internal audit reviews and whistleblower hotlines are detective controls focused on the human and governance dimensions of risk. An internal audit function that examines compliance with policies and controls will detect deviations — employees sharing passwords, approval processes being bypassed, access permissions that have not been updated when employees changed roles. A whistleblower hotline provides a channel for employees who observe misconduct to report it without fear of retaliation, detecting problems that internal audit might not find because they are being actively concealed.

Corrective Controls

A corrective control is designed to fix a problem after it has been identified, restore normal operations, and reduce the likelihood of recurrence. Corrective controls acknowledge that prevention and detection are imperfect — that some risk events will occur and will be detected, and that the firm must be prepared to respond effectively when they do.

Incident response plans are corrective controls. A documented, tested plan that specifies who does what, what decisions need to be made, how to communicate with affected customers and regulators, and how to restore compromised systems is the difference between a controlled, effective response and a chaotic scramble that compounds the damage.

An incident response plan that sits in a document repository and is never rehearsed is not a functioning corrective control — the corrective capability comes from having practiced the response, not from having written it down.

Data backup and disaster recovery systems are corrective controls that enable a firm to restore its operations after a destructive event — a ransomware attack that encrypts production data, a hardware failure that destroys a data center, a catastrophic software error that corrupts a database. The quality of a backup and recovery system is measured not by whether backups exist but by how quickly they can be restored to operational condition — a backup that takes three weeks to restore does not provide the recovery capability that a business requiring near-continuous availability needs. Recovery time objectives and recovery point objectives — the maximum acceptable time to restore service and the maximum acceptable data loss — are the metrics that define what the recovery system must be able to achieve.

Patch management is a corrective control that addresses identified vulnerabilities in software and systems. When a security vulnerability is discovered in software that a FinTech firm uses, the firm must assess the risk, obtain the vendor's fix or develop a workaround, test it, and deploy it to production — ideally before the vulnerability is exploited. Effective patch management is a continuous process, not an occasional activity, because new vulnerabilities are discovered constantly and the window between discovery and exploitation by attackers is often very short.

Customer notification and remediation procedures are corrective controls that address the harm done to customers when a risk event affects them. A payment platform that detects fraudulent transactions on customer accounts must notify those customers, reverse the fraudulent charges, and restore their access — all quickly enough to limit the harm and maintain trust. State breach notification laws and the GLBA Safeguards Rule impose legal requirements on the timing and content of customer notifications following data breaches; the corrective control program must be designed to meet those requirements as a minimum standard.

Root cause analysis — systematically investigating why a control failure or risk event occurred, identifying the underlying causes, and implementing changes to prevent recurrence — is a corrective control focused on institutional learning. An organization that treats each incident as an isolated event and does not investigate its causes will see the same types of incidents recur. An organization that conducts rigorous root cause analysis after significant incidents and redesigns processes or controls based on what it learns will gradually reduce its risk exposure over time.

The Security-Convenience Trade-off in Control Design

Every internal control involves a trade-off between security and usability. Convenience and speed are often at odds with security measures. The most secure system would prevent all unauthorized access — and all access. The most usable system would eliminate all friction — and all protection. Practical control design requires finding the point on this spectrum that provides adequate security for the risk being addressed without imposing costs — in time, friction, and complexity — that cause users to circumvent the control or that damage the business's competitive position.

This trade-off is most visible in authentication design. Multi-factor authentication substantially reduces the risk of unauthorized account access, but it adds steps to the login process that some users find cumbersome and that increase the volume of customer support requests from users who lose access to their second factor. A FinTech firm designing its authentication controls must weigh the security benefit of stronger authentication against the impact on user experience and the competitive risk of being more cumbersome to use than alternatives.

The same trade-off appears throughout the control landscape. Transaction velocity limits that prevent large-scale fraud also occasionally block legitimate transactions. Audit log review that detects unauthorized access also requires security staff time that could be spent on other activities. Segregation of duties that prevents a single employee from completing a fraudulent transaction also slows down legitimate transactions that require approval from a second party. Recognizing these trade-offs explicitly — rather than treating security controls as costless — is part of professional risk management.

Enterprise Resource Planning systems, commonly known as ERP systems, represent an attempt to address the security-usability trade-off by integrating security controls directly into the operational processes they protect. An ERP system is an example of a system that merges high security and data usability. Rather than adding security as a layer on top of operational processes, ERP systems build access controls, audit trails, segregation of duties, and data governance into the core of how business functions are performed. Employees execute their workflows within the system, which enforces control requirements automatically — generating audit logs, routing approvals to the right people, preventing users from accessing data outside their authorized scope — without requiring them to perform separate security procedures. This integration reduces friction while maintaining control effectiveness, which is why ERP systems are widely used in financial services for high-risk processes like accounting, procurement, and financial reporting.

Three Difficulties for Regulators Dealing with FinTech Fraud

Regulators face a specific set of challenges in addressing fraud in the FinTech sector that are distinct from the challenges of regulating traditional financial institutions. Three difficulties are particularly significant and recurrent.

The first is acting quickly enough to prevent FinTech fraud. FinTech transactions happen at machine speed across digital channels that span jurisdictions. A fraudster who gains unauthorized access to a payment platform account can initiate and complete transactions in seconds — far faster than any regulatory intervention can occur. By the time a regulator identifies a pattern of fraudulent activity and moves to act, the damage may already be done and the perpetrators may have moved funds through multiple accounts and across international borders. The speed advantage that makes FinTech valuable to consumers also makes the harm from fraud significantly harder to prevent through regulatory action.

The second difficulty is identifying the fraud. FinTech fraud is often invisible in ways that traditional financial fraud is not. A fraudster who opens thousands of synthetic identity accounts — accounts created using combinations of real and fabricated personal information — can operate for months before the fraud becomes detectable through patterns that emerge only in aggregate. A bad actor who exploits a lending platform's credit scoring model by feeding it manipulated data may not trigger fraud alerts designed to catch individual unauthorized transactions. The complexity and scale of FinTech operations creates fraud patterns that are difficult to distinguish from legitimate activity without sophisticated analytical capabilities that smaller regulators may not have.

The third difficulty is the lack of investigators with the technical expertise to understand and investigate FinTech fraud effectively. Traditional financial fraud investigation relies on examining paper trails, interviewing witnesses, and tracing the flow of funds through bank accounts. FinTech fraud investigation requires understanding blockchain transaction graphs, API exploitation techniques, synthetic identity patterns in large datasets, and the technical architecture of the systems being exploited. The pool of investigators with both financial fraud expertise and deep FinTech technical knowledge is small, and both industry and government compete for it. Regulatory agencies that cannot hire and retain technically sophisticated investigators will be structurally slower to detect and respond to FinTech fraud than the sector's pace of evolution requires.

When Controls Fail: Real-World FinTech Failures

Abstract risk management frameworks become concrete when examined against actual failures. Two of the most significant FinTech collapses of recent years — FTX and Wirecard — illustrate what happens when internal controls are absent, ignored, or actively circumvented. Both cases are worth examining in detail, because they demonstrate the relationship between control failures and catastrophic outcomes in ways that no hypothetical example can.

FTX: The Absence of Controls

FTX was the world's second-largest cryptocurrency exchange when it collapsed in November 2022, filing for bankruptcy within days of reports that customer funds had been misappropriated. The collapse wiped out approximately $8 billion in customer funds and resulted in the criminal conviction of founder Sam Bankman-Fried on seven counts of fraud and conspiracy.

The FTX failure was not primarily a market risk event or a cybersecurity incident — it was a control failure of comprehensive proportions. FTX had no independent board of directors. It had no chief risk officer. It had no internal audit function. Its accounting was conducted with QuickBooks, accounting software designed for small businesses, rather than systems appropriate for an exchange processing billions of dollars in daily transactions. Customer funds were commingled with trading funds at Alameda Research, FTX's affiliated trading firm, in direct violation of the basic financial services principle that customer assets must be segregated from firm assets. The person responsible for FTX's risk management was Bankman-Fried's romantic partner with no formal risk management qualifications.

From a controls perspective, FTX lacked preventative controls that would have stopped the commingling of customer funds — segregation of duties would have required that the decision to transfer customer assets to Alameda be made and approved by multiple independent parties, not by a single founder with no oversight. It lacked detective controls that would have identified the misappropriation — independent audit, reconciliation of customer balances against exchange holdings, and board-level financial review would all have revealed the gap between what FTX owed customers and what it actually held. And it lacked corrective controls that would have triggered remediation — a functioning compliance function with escalation authority, access to regulatory counsel, and independence from the founder would have identified and escalated the problem before it became a catastrophic fraud.

FTX's failure is an extreme case, but it illustrates a pattern that appears in less dramatic form across many FinTech startups: the belief that controls can be deferred until the firm reaches a certain scale, combined with a founder culture that treats oversight as an obstacle to speed. The specific controls whose absence enabled FTX's fraud — board independence, segregation of customer and firm assets, independent audit, professional risk management — are not optional luxuries for established firms. They are foundational to operating a financial services business with customers' money.

Wirecard: When Controls Are Present but Captured

Wirecard was a German payment processing firm that grew into a major FinTech company and was briefly included in the DAX, Germany's blue-chip stock index. In June 2020, the company disclosed that approximately 1.9 billion euros in cash that auditors had been told was held in escrow accounts in the Philippines did not appear to exist. The company filed for insolvency days later — the largest insolvency in German corporate history at the time.

Wirecard's control failure differs instructively from FTX's. Unlike FTX, Wirecard had formal control structures: it had external auditors, a supervisory board, and regulatory oversight. But those controls were compromised at multiple points. The external auditor — EY — signed off on Wirecard's financial statements for years without successfully verifying the existence of the cash that made up a substantial portion of the company's reported assets. German financial regulators, responding to mounting journalistic investigations of the company's accounting, initially investigated the journalists rather than the company. Short sellers who identified problems were threatened with legal action.

The Wirecard case illustrates that the formal existence of controls does not guarantee their effectiveness. Detective controls — including external audit — can fail if auditors do not exercise sufficient professional skepticism, if management actively provides false documentation, or if the institutional pressure to maintain a relationship with a major client suppresses the independence that makes audit valuable. It also illustrates the importance of external accountability mechanisms beyond the firm's own control program: the Financial Times reporting that ultimately forced the crisis into the open was itself a form of detective control — investigative journalism serving the function that the formal audit and regulatory oversight processes had failed to perform.

The Qualitative Risk Assessment in Practice

Understanding what a well-executed qualitative risk assessment looks like in practice brings the framework developed in this chapter to life.

A qualitative risk assessment is a structured document that captures the firm's current understanding of its risk landscape. Each row in the template represents one identified risk and contains: the threat name; the threat origin (internal, external, natural, accidental, or malicious); a brief description of what the threat is and how it could materialize; the business impact — what harm would result if the risk event occurred; the recommended corrective action — what control or response would mitigate the risk; and the likelihood, impact, and overall risk level scores.

The corrective action field is particularly important to understand precisely. In this context, corrective action refers to the recommended response to the risk — the control that should be implemented to prevent, detect, or correct it. This is not limited to corrective controls in the three-type taxonomy. The recommended action for a high-likelihood phishing risk might be a preventative control — mandatory phishing awareness training and email filtering. The recommended action for a medium-likelihood data breach risk might be a detective control — real-time intrusion detection and anomaly monitoring. The corrective action field captures whatever type of control is most appropriate for the specific risk.

A complete risk assessment spanning 20 risks should cover a range of threat origins and risk types. Technical risks — phishing, ransomware, API vulnerabilities, system outages — are important and should be included. But non-technical risks are equally important: regulatory change, key person dependency, vendor failure, reputational damage from a customer service failure, and the risk that the firm's fraud controls fail to keep pace with evolving attack techniques. An assessment that covers only technical risks has not reflected the full scope of what the firm faces.

The risk level scores — likelihood and impact, each rated on a qualitative scale, combined to produce an overall risk level — are the mechanism that distinguishes the assessment from a simple list of concerns. High-likelihood, high-impact risks require immediate, substantial control investment. Low-likelihood, low-impact risks may be accepted or addressed with minimal controls. The scoring process forces prioritization — the recognition that resources for risk management are finite and must be directed where they provide the most value.

Key Takeaways

Overall risk is generally higher in FinTech firms than in traditional financial firms across all major risk categories — market, credit, operational, and business risk. This elevation reflects FinTech's digital-first architecture, rapid scaling culture, reliance on third-party infrastructure, and the relative immaturity of most FinTech firms' risk management and compliance functions. FinTech startups, in particular, tend to worry less about security than traditional banks — a documented pattern that creates vulnerability gaps that must be consciously addressed as firms mature.

The major types of risk facing FinTech firms include: operational risk from system failures, process gaps, and third-party dependencies; market risk amplified by cryptocurrency volatility and algorithmic trading; credit risk from serving underbanked populations with less-tested scoring models; business risk from regulatory change and dependency on specific partnerships; IT and technology risk from the total dependence on digital infrastructure; compliance risk from operating in fragmented and evolving regulatory environments; and reputational risk from the high visibility of digital-first failures.

The risk management process moves through identifying risks by threat origin (internal, external, natural, accidental, malicious), assessing them on likelihood and impact to produce a risk level, selecting responses (mitigation, transfer, acceptance, avoidance), implementing controls, and monitoring effectiveness. A complete risk inventory should span both technical and non-technical risk categories, covering system vulnerabilities alongside regulatory, operational, and human risks.

The three types of internal controls are distinguished by their timing and purpose. Preventative controls stop risk events before they occur —

examples include multi-factor authentication, least-privilege access controls, segregation of duties, input validation, employee background checks, and encryption. Detective controls identify that a risk event has occurred or is occurring — examples include real-time transaction monitoring, audit logs, automated reconciliation, intrusion detection systems, penetration testing, internal audit reviews, and whistleblower hotlines. Corrective controls fix problems after detection and prevent recurrence — examples include incident response plans, data backup and disaster recovery systems, patch management, customer notification and remediation procedures, and root cause analysis. A strong control program includes all three types working together.

Convenience and speed are often at odds with security measures. Every internal control involves a trade-off between security and usability, and effective control design requires finding the level of friction that provides adequate security without unacceptably degrading the user experience or the firm's competitive position. Enterprise Resource Planning systems are a notable example of a system that merges high security and data usability by integrating control requirements directly into operational workflows.

Regulators face three specific difficulties in addressing FinTech fraud: acting quickly enough to prevent it, given the speed of digital transactions; identifying it, given the complexity and scale of FinTech operations and the sophistication of fraud techniques; and the lack of investigators with the technical expertise to understand and pursue FinTech-specific fraud patterns.

The FTX collapse illustrates what happens when all three types of internal controls are absent: customer funds were commingled with firm assets without the segregation of duties and independent oversight that would have prevented it, detected it, or triggered corrective action before catastrophic harm occurred. The Wirecard case illustrates that formal control structures can fail when they are not executed with independence and professional skepticism — the existence of an audit function does not guarantee effective detection if the auditors are not exercising the skepticism their role requires.

Chapter 12: Corporate Governance, AML, KYC, and Regulatory Compliance

Introduction

Chapter 11 built the risk management framework and introduced the three types of internal controls. Controls, however, are only as effective as the institutional structures that require them, monitor them, and hold people accountable when they fail. A well-designed preventative control that no one enforces is not a functioning control. A detective control that generates alerts no one reviews is not providing detection value. A corrective control that requires independent approval from a board that does not actually scrutinize management decisions is not providing real oversight.

Corporate governance is the system of structures, rules, and practices by which a firm is directed and controlled. It determines who makes decisions, who oversees those decisions, and who is accountable when decisions go wrong. In financial services, governance is not merely an organizational design question — it is a regulatory requirement and a fundamental condition of operating with other people's money.

This chapter covers the governance and compliance layer that gives internal controls institutional force. It begins with what corporate governance means in the FinTech context and why it matters. It examines the compliance function — what it does, how it is staffed, and how it evolves as a firm grows. It covers anti-money laundering requirements and know-your-customer obligations in the depth that FinTech professionals need to understand them. It addresses financial fraud, tax compliance, and the distinction between legal tax avoidance and illegal tax evasion. And it examines three instructive cases — Hammersmith and Fulham, Cathay Pacific, and Berkeley Coin — that illustrate what happens when governance and compliance fail in different but related ways.

Together, risk management and governance are inseparable in practice. Controls without governance are rules without enforcement. Governance without controls is oversight without mechanisms. Effective risk management requires both, and the two chapters in this pair are designed to be read together.

Corporate Governance in FinTech

Corporate governance is the framework of relationships, authorities, and accountability mechanisms that determines how a company is directed. In its most basic form, it answers three questions: who has the authority to make decisions, who oversees the exercise of that authority, and what happens when that authority is abused or exercised poorly.

In a traditional public company, the governance structure is layered and formalized. Shareholders elect a board of directors. The board hires and oversees executive management, approves major strategic decisions, and holds management accountable for performance. An audit committee of the board oversees financial reporting and the relationship with external auditors. A compensation committee oversees executive pay. A risk committee oversees the firm's risk management program. External auditors provide independent verification of financial statements. Regulators examine the firm's operations and compliance with applicable law. This multi-layered structure creates overlapping checks on management authority — multiple parties, each with different interests and information, scrutinizing what management is doing.

FinTech startups typically begin with none of this structure. A founding team of two or three people makes all decisions collectively. There is no board with independent directors — there may be a board consisting entirely of founders and early investors with a financial interest in the company's growth, not in conservative risk management. There is no audit committee because there is no audit. There is no compliance function because no one has been hired to perform it. These absences are not necessarily irresponsible in the very early stages — a five-person startup cannot afford a full governance apparatus. But the absence of governance structures that are appropriate for a small startup becomes dangerous when the firm grows, when it begins handling significant customer funds, and when the decisions being made have consequences that extend well beyond the founding team.

The governance gap — the period between when a FinTech firm is too small to warrant full governance structures and when it has built them — is one of the most significant sources of risk in the sector. During this gap, significant decisions are being made informally, significant customer funds are being handled without independent oversight, and significant compliance obligations are being managed by people who may not be qualified compliance professionals. The history of FinTech is substantially a history of what happens when firms scale their customer base without scaling their governance.

Board Structure and Independence

An effective board of directors provides oversight that management cannot provide for itself. The key word is independent. A board composed entirely of the company's founders and executives cannot provide meaningful oversight of those same founders and executives — it has no independence, no ability to ask hard questions from a disinterested position, and no incentive to challenge decisions that may be profitable in the short term but risky or improper in the longer term.

Board independence means having directors who are not employees, not major shareholders with a financial interest in the firm's short-term performance, and not personal friends or business associates of management who lack the willingness to challenge the people who appointed them. An independent director brings outside perspective, professional expertise, and the institutional position to say no — to reject an acquisition, question a financial practice, demand answers about a compliance failure, or require changes to executive compensation that creates perverse incentives.

For regulated financial institutions, board composition is not merely a governance best practice — it is a regulatory requirement. Bank regulators examine the composition and effectiveness of boards as part of routine examination. The FDIC, OCC, and Federal Reserve all have guidance on what constitutes effective board oversight. FINRA rules require broker-dealers to have governance structures that include meaningful compliance oversight. FinTech firms that operate under bank charters, broker-dealer registrations, or money transmitter licenses inherit regulatory expectations about governance that informal startup structures do not meet.

Audit committees play a particularly important governance role. An audit committee consisting of independent directors with financial expertise oversees the integrity of financial reporting, the relationship with external auditors, and the effectiveness of internal audit. When an audit committee is functioning properly, it is the mechanism through which the board learns whether management's financial representations are accurate and whether internal controls are actually working. When an audit committee is captured — dominated by directors who are too close to management to exercise independent judgment — financial fraud can persist for years before it becomes visible externally.

Fiduciary Duties

Directors and senior executives of financial firms owe fiduciary duties to the firm and, through the firm, to its customers and stakeholders. A fiduciary duty is a legal and ethical obligation to act in the best interests of another party — placing that party's interests above one's own. In the corporate context, fiduciary duties include the duty of care, which requires directors to make decisions with the care and diligence that a reasonably prudent person would exercise in similar circumstances, and the duty of loyalty, which prohibits self-dealing and requires directors to put the firm's interests ahead of their personal interests.

In financial services, fiduciary duties have a particular weight because the firm is handling other people's money. A FinTech lending platform whose directors approve policies designed to maximize the firm's fee income at the expense of customer financial outcomes is likely violating their duty of loyalty. A FinTech firm whose executives approve the use of customer funds for purposes the customers did not authorize — as occurred at FTX — is committing a fundamental breach of fiduciary obligation that crosses from governance failure into fraud.

The distinction matters because it defines both the legal standard that governance structures must meet and the ethical standard that professionals who serve in governance roles must apply. Serving on a board is not a ceremonial role. It carries real legal and ethical obligations, and directors who fail to exercise genuine oversight can face personal liability for governance failures that occur on their watch.

The Compliance Function: From Startup to Scale

A compliance officer is a professional responsible for ensuring that a firm meets its legal and regulatory obligations. In a large bank, the Chief Compliance Officer leads a department that may include dozens of specialists covering different regulatory domains — AML compliance, consumer protection, securities law, data privacy, and more. In a FinTech startup, compliance may initially be performed by a founder or general counsel alongside other responsibilities, with no dedicated compliance staff at all.

The compliance function exists because the regulatory obligations that financial firms must meet are too complex, too numerous, and too consequential to be managed informally. A compliance officer who is genuinely independent — who can raise concerns to the board without going through the CEO, who has the authority to halt a product launch that creates unacceptable compliance risk, and who is not evaluated primarily on metrics that create pressure to approve things quickly — provides a check on operational and legal risk that no other function can substitute for.

The tension between compliance and growth is one of the defining dynamics of FinTech firms at the scaling stage. A compliance officer who identifies that a new product feature requires additional regulatory approval before launch is, in the short term, creating friction that slows revenue growth. A compliance team that flags a marketing practice as potentially misleading under consumer protection law is, from a product team's perspective, blocking a campaign. These tensions are real, and they play out in every FinTech firm that is simultaneously trying to grow rapidly and operate within a complex regulatory environment.

How those tensions are resolved is a governance question as much as a compliance question. A firm in which the compliance officer reports directly to the CEO — who has an obvious interest in resolving compliance questions in favor of growth — has a structurally compromised compliance function. A firm in which the compliance officer has a reporting line to the board's audit or risk committee, independent of the CEO, has a compliance function with real institutional independence. The reporting structure of the compliance function is itself a governance design decision with significant implications for whether compliance is genuinely effective or performative.

Compliance theater — the maintenance of formal compliance structures that satisfy regulatory appearances without providing genuine oversight — is a recurring pattern in FinTech failures. Firms that have a compliance officer on the organizational chart but have not given that officer the authority, resources, or institutional independence to actually function are not more compliant than firms with no compliance officer. They are simply better positioned to claim, after the fact, that they had a compliance program.

Regulation as a Balancing Act

Regulation in financial services is a balancing act between protecting businesses and protecting individuals. This is not a comfortable equilibrium — the interests being balanced are genuinely in tension, and reasonable people disagree about where the balance should be struck. Regulation that is too light leaves consumers exposed to fraud, discrimination, and the consequences of inadequate controls. Regulation that is too heavy imposes compliance costs that reduce competition, increase prices, and drive activity into less regulated channels.

The framing of regulation as a balancing act between protecting businesses and protecting individuals captures something important that simpler framings miss. It is not a balance between regulation and freedom — both regulated and unregulated markets involve tradeoffs of different kinds of freedom. It is not a balance between innovation and safety — well-designed regulation can enable innovation by providing clear rules that allow firms to operate confidently. It is specifically a balance between the interests of firms, which need regulatory clarity and manageable compliance costs to operate and grow, and the interests of individuals, who need protection from the harms that unregulated financial markets have historically produced.

What good and bad regulations have in common is that they both create opportunities for creative and innovative solutions. This is a counterintuitive but important observation. A regulation requiring that consumer financial data be portable — that customers can take their financial history to a new provider — creates a market for innovative services that compete on the quality of what they do with that data. A regulation requiring that AML programs be written and tested creates a market for compliance technology that automates and improves those programs. Even poorly designed regulations that create compliance burdens create markets for firms that find clever ways to meet those burdens more efficiently. Regulation shapes the environment in which innovation occurs; it does not simply suppress it.

This does not mean all regulation is equally good. Poorly designed regulations impose costs without producing proportionate benefits, create barriers to entry that protect incumbents rather than consumers, or generate perverse incentives that produce the opposite of their intended effects. The point is that regulation and innovation are not simply in opposition. The most innovative sectors of the economy operate under significant regulatory frameworks — pharmaceuticals, aviation, financial services — because those frameworks create the baseline of trust and safety that makes markets function.

Anti-Money Laundering: The Bank Secrecy Act Framework

Money laundering is the process of making the proceeds of criminal activity appear to come from a legitimate source. Drug trafficking, fraud, corruption, human trafficking, and terrorism all generate cash or financial assets that cannot be spent or invested openly without revealing their criminal origin. Money laundering is the mechanism by which criminal proceeds are integrated into the legitimate economy — moved through a series of transactions designed to obscure the connection between the money and the crime that produced it.

The Bank Secrecy Act, enacted in 1970, is the foundational U.S. anti-money laundering statute. It requires financial institutions to maintain records and file reports that help law enforcement detect, investigate, and prosecute money laundering and related financial crimes. FinCEN — the Financial Crimes Enforcement Network, a bureau of the U.S. Treasury — administers the Bank Secrecy Act and receives the reports that financial institutions are required to file.

The Bank Secrecy Act's reach has expanded significantly since 1970, both through amendments and through FinCEN's regulatory authority. The USA PATRIOT Act of 2001 strengthened AML requirements substantially, adding customer identification program requirements and enhanced due diligence obligations for certain categories of customers. The Anti-Money Laundering Act of 2020 — the most significant reform of the Bank Secrecy Act since the PATRIOT Act — modernized the AML framework, required FinCEN to establish beneficial ownership registries, and updated information-sharing provisions between financial institutions and law enforcement.

Who Must Comply

The Bank Secrecy Act applies to financial institutions, a category that includes banks, credit unions, broker-dealers, futures commission merchants, insurance companies — and money services businesses. Money services businesses, or MSBs, is the category that captures most FinTech firms. An MSB is defined to include currency dealers or exchangers, check cashers, issuers of traveler's checks or money orders, sellers or redeemers of stored value, money transmitters, and — critically for the FinTech sector — virtual currency exchangers and administrators.

A FinTech firm that moves money — a payment app, a digital wallet, a crypto exchange, a peer-to-peer lending platform that handles fund transfers — is almost certainly an MSB and must comply with Bank Secrecy Act requirements. This means registering with FinCEN, implementing a written AML program, conducting customer due diligence, filing currency transaction reports and suspicious activity reports, and complying with the travel rule for certain transfers. The table below summarizes the key requirements and their application to FinTech firms.

Requirement	What It Requires	Who Enforces It	FinTech Application

Bank Secrecy Act Registration	MSBs must register with FinCEN before operating	FinCEN	Payment apps, crypto exchanges, and digital wallets that move money must register as MSBs
Written AML Program	Formal, documented AML compliance program with designated compliance officer, internal controls, independent testing, and ongoing training	FinCEN, federal banking regulators	FinTech firms must have a written program — informal procedures do not satisfy the requirement
Customer Identification Program (KYC)	Verify customer identity at account opening; maintain records; screen against government watch lists	FinCEN; enforced by banking regulators for bank-chartered firms	Identity verification at onboarding; ID document collection; OFAC screening
Customer Due Diligence (CDD)	Understand the nature and purpose of customer relationships; collect beneficial ownership information for business accounts	FinCEN	Risk-based profiling of customers; enhanced scrutiny for higher-risk relationships
Currency Transaction Reports (CTRs)	File CTR for cash transactions exceeding $10,000	FinCEN	Primarily relevant to FinTech firms handling physical cash; less common in digital-first platforms
Suspicious Activity Reports (SARs)	File SAR when a transaction involves $5,000 or more and the firm knows,	FinCEN	Automated transaction monitoring systems generate

	suspects, or has reason to suspect illegal activity		SAR referrals; compliance team reviews and files
Travel Rule	Financial institutions must transmit identifying information about the originator and beneficiary of wire transfers above $3,000	FinCEN	Applies to crypto transfers above threshold; creates compliance challenges for transfers to unhosted wallets
Recordkeeping	Maintain records of transactions and customer information for five years	FinCEN	Digital records must be preserved and retrievable; deletion policies must account for legal retention requirements

The Written AML Program

The Bank Secrecy Act requires covered financial institutions to implement a written AML program that is reasonably designed to prevent the firm from being used to launder money or finance terrorism. A written program is not sufficient on its own — it must be implemented, tested, and updated as the firm's business and the regulatory environment evolve.

An effective AML program has four minimum elements: internal policies, procedures, and controls; a designated compliance officer with day-to-day responsibility for the program; ongoing training for relevant employees; and independent testing of the program's effectiveness. These are minimum requirements, not a complete description of what an effective program looks like. A large payment platform processing millions of transactions daily needs far more sophisticated AML controls than these four elements alone suggest — including automated transaction monitoring systems, risk-based customer segmentation, and a professional compliance team capable of reviewing and acting on the alerts those systems generate.

Suspicious activity reports — SARs — are the primary mechanism through which financial institutions report potential money laundering to law enforcement. A SAR must be filed when a transaction involves $5,000 or more and the financial institution knows, suspects, or has reason to suspect that the transaction involves funds from illegal activity, is designed to evade reporting requirements, lacks a lawful purpose, or involves a customer who is engaged in illegal activity. The SAR reporting obligation applies even when the institution is uncertain — the standard is reason to suspect, not proof of illegal activity.

SAR filing is protected by a statutory safe harbor: a financial institution that files a SAR in good faith cannot be sued by the customer who is the subject of the report. Institutions are also prohibited from tipping off the customer — informing a customer that a SAR has been filed or is being considered is itself a federal crime. This creates a distinctive compliance situation in which the institution must simultaneously maintain its customer relationship and report suspicious activity about that customer to law enforcement without the customer's knowledge.

Transaction Monitoring

The practical heart of an AML program is transaction monitoring — the systematic review of customer transactions to identify patterns that may indicate money laundering or other financial crime. Manual review of every transaction is not feasible at the scale at which FinTech platforms operate. Automated transaction monitoring systems analyze transactions in real time or near-real time, comparing each transaction against rules and behavioral models designed to identify suspicious patterns.

Effective transaction monitoring is harder than it sounds. Rules-based systems — flag any transaction above threshold X, flag any account that receives more than Y deposits in Z days — generate large numbers of false positives, in which legitimate transactions are flagged as suspicious. These false positives consume compliance team time and resources and, if not managed carefully, lead to alert fatigue in which reviewers process alerts quickly without adequate scrutiny. Machine learning-based monitoring systems improve on simple rules by identifying anomalous patterns relative to a customer's established behavior, reducing false positive rates while maintaining detection effectiveness. But these systems require significant investment, produce their own errors, and require ongoing calibration as criminal techniques evolve.

The compliance cost of transaction monitoring is one of the reasons that AML compliance is a meaningful competitive consideration in FinTech. A payment platform with millions of customers and billions of transactions must invest substantially in monitoring infrastructure to meet its SAR filing obligations. A startup that has not made that investment may be meeting the technical letter of the written program requirement without having the monitoring capability that effective AML compliance actually requires.

Know Your Customer: Identity, Due Diligence, and Fraud Prevention

Know Your Customer — KYC — is the collective term for the obligations that financial institutions have to verify the identity of their customers and understand the nature of those customers' activities. KYC requirements exist for two interconnected reasons: to help governments track illegal money by ensuring that financial institutions know who their customers are and can provide that information to law enforcement, and to ensure that institutions report clients' illegal activities if necessary by having sufficient information about customer relationships to identify when those relationships are being used for criminal purposes.

KYC is not simply a compliance formality. It is a substantive risk management tool. A financial institution that does not know who its customers are cannot identify when a customer relationship is being used for money laundering, fraud, or sanctions evasion. It cannot fulfill its SAR filing obligations because it lacks the information to identify suspicious activity. And it cannot protect itself from the legal, regulatory, and reputational consequences of facilitating financial crime.

Customer Identification: The Foundation of KYC

The Customer Identification Program, or CIP, requirement establishes the baseline KYC obligation: before opening an account, a financial institution must collect and verify certain identifying information about the customer. For individual customers, this means collecting name, date of birth, address, and an identification number — typically a Social Security number for U.S. persons or a passport number for foreign nationals. For business customers, it means collecting the legal name, address, and tax identification number, and — under FinCEN's beneficial ownership rule — identifying the natural persons who ultimately own or control the business.

Verification means confirming that the information provided is accurate and that the customer is who they claim to be. Documentary verification involves reviewing government-issued identification documents — a driver's license, passport, or other official ID. Non-documentary verification uses other means to confirm identity — database checks against credit bureau records, public records, or other data sources. For digital-first FinTech firms that onboard customers entirely through mobile apps or websites, remote identity verification using document scanning, facial recognition, and liveness detection has become the primary verification method, raising its own questions about the reliability of automated verification systems and their differential performance across demographic groups.

OFAC screening — checking customer names and identifying information against the Office of Foreign Assets Control's list of sanctioned individuals, entities, and countries — is a required element of the customer identification process. A financial institution that opens an account for a sanctioned individual or entity is not merely violating AML rules; it is potentially violating U.S. sanctions law, which carries severe civil and criminal penalties. OFAC screening must occur at onboarding and on an ongoing basis, because the sanctions lists are updated frequently as new individuals and entities are designated.

Customer Due Diligence and Enhanced Due Diligence

Customer due diligence — CDD — goes beyond identification to require financial institutions to understand the nature and purpose of customer relationships and to use that understanding to assess the risk that the customer relationship may be used for money laundering or financial crime. A basic CDD program involves collecting information about what the customer does, why they are opening an account, and what types of transactions they expect to conduct. This baseline information establishes the expected pattern of the customer's behavior — the standard against which future transactions are compared to identify anomalies.

Enhanced due diligence — EDD — applies to higher-risk customer relationships. Politically exposed persons — PEPs — are individuals who hold or have held prominent public positions, including heads of state, senior government officials, military officers, senior executives of state-owned enterprises, and their close family members and associates. PEPs are considered higher-risk because their positions create elevated exposure to bribery and corruption. A FinTech lending platform that onboards a senior foreign government official must apply EDD — collecting more information, conducting more thorough background screening, obtaining senior management approval for the relationship, and monitoring it more closely on an ongoing basis.

Other categories that trigger enhanced due diligence include customers located in high-risk jurisdictions identified by FinCEN or the Financial Action Task Force, customers in industries associated with elevated money laundering risk, and customers whose transaction activity is inconsistent with their stated purpose. The risk-based approach to KYC — calibrating the level of due diligence to the assessed level of risk — is the standard framework endorsed by U.S. regulators and international bodies. It allows institutions to focus their compliance resources where the risk of financial crime is highest rather than applying uniform scrutiny to every customer relationship regardless of risk level.

KYC in the FinTech Context

KYC presents distinctive challenges for FinTech firms. The frictionless onboarding experience that FinTech products are designed to deliver — opening an account in minutes with a few taps — is in direct tension with the thorough identity verification and due diligence that KYC requires. The fastest possible onboarding experience and the most thorough possible identity verification are not the same thing, and FinTech firms must design their onboarding processes to balance these competing objectives.

The emergence of digital identity verification technology has reduced but not eliminated this tension. Automated document scanning can verify the authenticity of identity documents in seconds. Facial recognition can compare a selfie to the photo on an identity document with high accuracy. Liveness detection can confirm that the person submitting the verification is physically present rather than submitting a photograph. These technologies make remote KYC faster and more reliable than manual review — but they are not perfect, and their error rates vary significantly across demographic groups, creating both compliance risk and fairness concerns.

Crypto exchanges face particularly challenging KYC requirements because the pseudonymous nature of blockchain transactions — the feature that makes crypto attractive to privacy-conscious users — conflicts directly with the identity verification requirements that AML law imposes. A crypto exchange that allows users to trade without identity verification is not merely offering privacy; it is facilitating the use of its platform for money laundering and is violating its Bank Secrecy Act obligations. The major exchanges have responded by implementing robust KYC programs, but smaller platforms and decentralized exchanges continue to operate in ways that create regulatory and legal exposure.

Financial Fraud, Tax Compliance, and Legal Boundaries

Financial fraud takes many forms in the FinTech context, and understanding the legal boundaries that define fraud — versus aggressive but legal business practices — is essential for compliance professionals and FinTech practitioners. Two distinctions are particularly important and are frequently misunderstood.

Tax Avoidance Versus Tax Evasion

Legal tax avoidance is not a crime. Tax avoidance refers to the use of legal means to reduce a tax obligation — taking advantage of deductions, credits, timing strategies, and structural choices that the tax code explicitly permits. Every business and individual that takes a mortgage interest deduction, contributes to a retirement account, or structures a transaction to achieve favorable tax treatment is engaging in tax avoidance. The tax code creates these opportunities deliberately, as a matter of legislative policy, and using them as intended is not illegal.

Tax evasion is a crime. Tax evasion involves deliberately concealing income, falsifying records, or otherwise misrepresenting one's tax obligations to pay less than the law requires. A FinTech firm that keeps two sets of books — one for tax reporting and one reflecting actual income — is committing tax evasion. An individual who fails to report cryptocurrency gains as taxable income is potentially committing tax evasion, not merely engaging in aggressive tax avoidance.

The distinction matters practically because FinTech firms encounter it in several contexts. Cryptocurrency investors frequently attempt to minimize their tax obligations through strategies ranging from clearly legal — tax-loss harvesting, holding assets for long-term capital gains treatment — to clearly illegal — simply not reporting gains at all. Most cryptocurrency investors do not successfully avoid capital gains taxes. The IRS has made crypto tax reporting a priority, requiring taxpayers to disclose cryptocurrency transactions on their tax returns and working with exchanges to obtain customer transaction records through summons and court orders. An exchange that facilitates tax evasion by its customers — by providing false records or enabling transactions specifically designed to obscure taxable events — has its own legal exposure beyond its customers' tax liabilities.

For FinTech firms designing products, the tax avoidance versus evasion distinction shapes what product features are permissible. A robo-advisor that automatically executes tax-loss harvesting strategies for customers is offering a tax avoidance service that is entirely legal. A crypto platform that markets its services based on the claim that customers can avoid paying taxes is potentially facilitating tax evasion and creating liability for itself alongside its customers.

Common Forms of Financial Fraud in FinTech

Financial fraud in the FinTech sector takes several recurring forms that compliance programs must be designed to detect and prevent. Identity fraud — using stolen or fabricated identity information to open accounts and access credit — is the most common form and the primary target of KYC programs. Synthetic identity fraud, which involves creating fictitious identities by combining real and fabricated information, is a particularly sophisticated variant that traditional fraud detection systems struggle to identify because each individual piece of information may check out even though the identity as a whole is fictitious.

Account takeover fraud occurs when an attacker gains access to a legitimate customer's account and uses it to initiate unauthorized transactions. The combination of stolen credentials, social engineering, and SIM-swapping attacks has made account takeover a significant source of loss for payment platforms, lending apps, and investment platforms. The controls against account takeover — multi-factor authentication, device recognition, behavioral biometrics, and anomaly detection — overlap substantially with cybersecurity controls, illustrating the connection between the risk management framework of Chapter 11 and the fraud compliance framework of this chapter.

First-party fraud occurs when a customer deliberately misuses a financial product — applying for a loan with no intention of repaying it, disputing legitimate transactions as unauthorized, or exploiting promotional offers in ways the firm did not intend. First-party fraud is the most common type of lending platform fraud and is particularly challenging to detect because the fraudulent customer is using their own real identity and the transactions appear legitimate until the customer fails to repay or successfully disputes a charge.

Wire fraud and securities fraud — the federal criminal statutes most commonly applied to financial misconduct — carry substantial criminal penalties and are tools prosecutors use broadly in FinTech fraud cases. The wire fraud statute is particularly expansive: any fraudulent scheme that uses electronic communications in interstate commerce can be charged as wire fraud, which means that virtually any financial fraud involving digital systems or the internet has wire fraud exposure. The FinTech cases that have resulted in the most significant criminal prosecutions — FTX, Celsius, BitConnect — were charged primarily under wire fraud and securities fraud statutes rather than specialized financial regulation.

Governance Failures in Practice: Three Cases

Abstract governance principles become concrete when examined through real failures. Three cases illustrate different dimensions of how governance and compliance failures enable financial harm: the Hammersmith and Fulham derivatives case, which established foundational principles about authorization and governance accountability; the Cathay Pacific fuel hedging case, which illustrates how inadequate governance over internal risk-taking can produce losses even in favorable market conditions; and the Berkeley Coin case, which illustrates how regulatory arbitrage through ICO structures has been used to attempt to avoid securities regulation.

Hammersmith and Fulham: Unauthorized Trading and Governance Accountability

In the late 1980s, the London Borough of Hammersmith and Fulham — a municipal local government in the United Kingdom — entered into a large number of interest rate swap contracts with major banks. Interest rate swaps are derivative instruments that allow counterparties to exchange fixed-rate and floating-rate interest payment obligations, typically used to manage exposure to interest rate movements. Hammersmith and Fulham's treasurer, seeking to generate returns for the borough, entered into an enormous volume of these contracts — far beyond anything that could be characterized as hedging the borough's existing interest rate exposure.

When interest rates moved against the borough's positions, Hammersmith and Fulham faced massive losses on the swap contracts — losses that would have fallen on local taxpayers. The borough's management argued that the contracts should be voided on the grounds that the local authority lacked the legal authority — the capacity — to enter into speculative derivative transactions. English local authorities are creatures of statute, possessing only the powers that legislation explicitly grants them. Speculative derivatives trading was not among those powers.

The case proceeded through the English courts. The House of Lords, England's highest court at the time, ultimately decided that the trades were not authorized and the debts should be voided. The court held that the interest rate swap contracts were ultra vires — beyond the legal capacity of the local authority — and therefore unenforceable. The banks that had entered into the contracts with Hammersmith and Fulham lost their claims entirely, despite the fact that the contracts were commercially standard and had been executed in the normal course of business.

The Hammersmith and Fulham case established several governance principles that remain relevant. First, authorization matters — the person or entity entering into a financial commitment must have the actual legal authority to do so, and the counterparty bears responsibility for verifying that authority. Second, governance failures can have consequences that fall on parties who acted in good faith — the banks that entered into the swaps had no reason to suspect they were dealing with an entity that lacked capacity, yet they bore the loss. Third, speculation disguised as risk management is a governance failure pattern that internal controls must be designed to detect — a treasurer who is taking speculative positions while characterizing them as hedging is circumventing the risk management function in a way that only independent oversight can catch.

Cathay Pacific: Internal Speculators and the Limits of Risk Policy

Cathay Pacific, the Hong Kong-based airline, provides a case study in how internal actors can use risk management instruments for speculation in ways that produce losses that should not have been possible. Airlines face significant exposure to fuel prices — jet fuel is one of their largest operating costs, and price volatility creates planning uncertainty. To manage this exposure, airlines commonly hedge their fuel costs by purchasing futures contracts or options that lock in future fuel prices, providing cost certainty even if market prices move unfavorably.

Cathay Pacific's fuel hedging program was designed for exactly this purpose — to reduce the airline's exposure to fuel price volatility by locking in prices for a portion of its anticipated fuel needs. However, internal financial personnel within the airline went beyond the hedging mandate and entered into positions that were effectively speculative bets on the direction of oil prices — specifically, positions that would profit if oil prices rose.

When oil prices dropped significantly — a development that should have been beneficial for an airline that uses fuel as a major input cost — Cathay Pacific still lost money on its fuel positions because its internal speculators had bet that prices would rise. The hedges that should have protected the airline had been replaced or supplemented by speculative positions that moved against the airline when the market moved in a direction that should have been favorable.

The Cathay Pacific case illustrates a governance failure that is distinct from simple fraud. The internal personnel who took the speculative positions were not necessarily acting with criminal intent — they were making risk-taking decisions that exceeded their mandate, in a governance environment that did not adequately monitor or constrain their authority. The failure was not the absence of a risk management policy; it was the absence of controls sufficient to ensure that the policy was actually being followed. An independent risk management function with real authority to monitor and limit trading activity, reporting to a board risk committee rather than to the same management that the traders reported to, could have identified and stopped the position-taking before it produced significant losses.

Berkeley Coin: Regulatory Arbitrage Through ICO Structure

The Berkeley Coin case illustrates how ICO structures have been used to attempt to sidestep securities regulation — and why that attempt is legally precarious. Berkeley, California proposed a municipal bond alternative in which the city would issue digital tokens representing a claim on future municipal revenues, selling them to investors through an initial coin offering rather than through the traditional municipal bond market.

A key aspect of the Berkeley Coin's implementation strategy was that it would be structured as an ICO and therefore could avoid securities regulations. The theory was that by structuring the offering as a token sale rather than a traditional securities offering, the city could bypass the disclosure requirements, registration obligations, and investor protection rules that apply to conventional securities offerings. This would allow the city to access capital markets more cheaply and quickly than traditional municipal bond issuance permits.

The regulatory problem with this theory is straightforward: the Howey Test, not the label applied to an instrument, determines whether something is a security. A digital token that represents an investment of money in a common enterprise with an expectation of profit from the efforts of others is a security regardless of whether it is called a coin, a token, or any other name. Berkeley Coin investors would be providing money to the city with an expectation of return — the essential characteristics of a securities investment — and the ICO structure would not change that analysis.

The broader significance of the Berkeley Coin case is what it illustrates about regulatory arbitrage more generally. The impulse to structure transactions to avoid regulatory requirements is not inherently improper — firms are generally entitled to organize their affairs in ways that minimize compliance burdens, within limits. But when the regulatory requirements being avoided are substantive investor protections — disclosure, registration, anti-fraud provisions — the arbitrage imposes real costs on investors who are not receiving the protections they would otherwise have. The SEC's position, consistently maintained, is that the economic substance of a transaction, not its form, determines its regulatory treatment.

Governance and Compliance as Competitive Advantage

The cases examined in this chapter — Hammersmith and Fulham, Cathay Pacific, Berkeley Coin, and the broader pattern of FinTech governance failures illustrated in earlier chapters — might suggest that governance and compliance are primarily constraints on what firms can do. They are that. But they are also something else: a source of competitive advantage for firms that build them well.

A FinTech firm with a genuinely effective compliance program has several advantages over a competitor that is cutting corners. It is less likely to face regulatory enforcement actions that are expensive, time-consuming, and reputationally damaging. It is less likely to experience the kind of control failure that produces customer harm and generates civil litigation. It is more likely to obtain banking partnerships, because banks conducting due diligence on FinTech partners scrutinize compliance programs carefully. And it is better positioned to scale into regulated markets — including international markets with demanding compliance requirements — because it has built the infrastructure that those markets require.

The firms that have dominated financial services over long periods — not the firms that grew fastest in their first five years, but the firms that built durable businesses — have generally been firms with strong governance and compliance cultures. The firms that grew fastest by cutting compliance corners have generally encountered their reckoning eventually: a regulatory enforcement action, a fraud that the controls would have caught, a governance failure that enabled misconduct the board should have prevented. Building governance and compliance infrastructure early, before it is strictly required, is not merely a compliance obligation. It is an investment in the durability of the business.

Key Takeaways

Corporate governance is the framework of structures, rules, and accountability mechanisms that determines how a firm is directed and controlled. In FinTech, the governance gap — the period between when a startup is too small to warrant full governance structures and when it has built them — is one of the most significant sources of risk in the sector. Board independence, fiduciary duties, and effective audit committees are not optional governance features for financial services firms; they are foundational requirements for operating with other people's money.

The compliance function provides the institutional structure through which regulatory obligations are met and internal controls are monitored. A compliance officer who reports independently to the board, has real authority to raise concerns, and is not evaluated primarily on metrics that create pressure to approve things quickly provides genuine oversight. Compliance theater — formal compliance structures that satisfy regulatory appearances without providing genuine oversight — is a recurring pattern in FinTech failures.

Regulation is a balancing act between protecting businesses and protecting individuals. What good and bad regulations have in common is that they both create opportunities for creative and innovative solutions. The FinTech sector has consistently demonstrated that regulatory requirements, rather than simply constraining innovation, shape the environment in which innovation occurs and create markets for firms that find better ways to meet compliance obligations.

The Bank Secrecy Act is the foundational U.S. anti-money laundering statute. It requires covered financial institutions — including money services businesses, which include most FinTech firms — to register with FinCEN, implement written AML programs, conduct customer due diligence, file currency transaction reports and suspicious activity reports, and comply with the travel rule. A suspicious activity report must be filed when a transaction of $5,000 or more gives the institution reason to suspect illegal activity. SAR filings are protected by statutory safe harbor, and tipping off customers about SAR filings is prohibited by federal law.

Know Your Customer requirements exist for two interconnected reasons: to help governments track illegal money by ensuring that financial institutions know who their customers are, and to ensure that institutions report clients' illegal activities when necessary by having sufficient information to identify criminal misuse of customer relationships. KYC includes customer identification programs, customer due diligence, enhanced due diligence for higher-risk relationships, and OFAC screening. The frictionless onboarding that FinTech products are designed to deliver is in direct tension with thorough identity verification, and managing that tension is a central compliance design challenge.

Legal tax avoidance is not a crime. It involves using legally permitted means — deductions, credits, timing strategies, and structural choices — to reduce tax obligations. Tax evasion is a crime. It involves deliberately concealing income or misrepresenting tax obligations. Most cryptocurrency investors do not avoid capital gains taxes; the IRS has made crypto tax reporting a priority, requiring disclosure of cryptocurrency transactions and obtaining exchange records through legal process.

The Hammersmith and Fulham case established that the trades were not authorized and the debts should be voided — the interest rate swap contracts were beyond the local authority's legal capacity and therefore unenforceable. The principle: the person or entity entering into a financial commitment must have actual legal authority to do so, and authorization must be verified, not assumed. The Cathay Pacific case illustrates how internal speculators who were hedging that prices would rise continued to lose money even when oil prices dropped, because the speculative positions they had taken exceeded their mandate and moved against the airline when the market moved in a favorable direction. The case illustrates that risk management policy without controls sufficient to enforce it provides limited protection. The Berkeley Coin case illustrates that a key aspect of the implementation strategy — structuring the offering as an ICO to avoid securities regulations — is legally precarious because the Howey Test applies to economic substance, not to the label applied to an instrument.

Chapter 13: Artificial Intelligence in FinTech — Opportunity, Bias, and Accountability

Introduction

A loan officer at a traditional bank makes a credit decision by reviewing an application, examining a credit report, considering the applicant's employment and income, and applying a combination of policy guidelines and professional judgment. The process takes time, is subject to human inconsistency, and reflects both the strengths and limitations of individual human judgment — including, as decades of research have documented, the biases that individual humans bring to consequential decisions.

An AI-driven credit decisioning system makes the same determination in milliseconds, incorporating hundreds or thousands of variables simultaneously, applying the same model to every applicant with perfect consistency, and drawing on patterns in training data that no human analyst could detect or process manually. The speed advantage is real. The consistency advantage is real. The scale advantage is real. And the potential for the model's outputs to reflect, embed, and amplify the biases present in its training data — at a scale and speed that human judgment never could — is equally real.

Artificial intelligence is simultaneously the most powerful tool available for expanding access to financial services and one of the most significant sources of new risk in the FinTech sector. It is making fraud detection faster and more accurate. It is enabling credit access for people that traditional models excluded. It is automating compliance functions that would otherwise require large, expensive human teams. And it is generating gender-biased credit decisions, enabling deepfake fraud at industrial scale, creating accountability gaps that existing regulatory frameworks were not designed to address, and raising fundamental questions about transparency, fairness, and who bears responsibility when an algorithm causes harm.

This chapter covers the full picture. It begins with what AI actually is in the FinTech context — distinguishing genuine AI from simpler automation and explaining the techniques most relevant to financial services. It maps the major applications of AI across the FinTech landscape. It examines algorithmic bias in depth, including what causes it, why it is hard to detect and correct, and what the Apple Card case reveals about how it manifests in practice. It covers deepfakes as an AI-enabled fraud vector. It examines RegTech — the use of AI in regulatory compliance itself. And it addresses the accountability question: what obligations do FinTech firms have when their AI systems cause harm, and what regulatory frameworks are emerging to enforce those obligations.

What Artificial Intelligence Is — and What It Is Not

The term artificial intelligence is used loosely in business contexts to describe a range of technologies that vary significantly in their sophistication and capabilities. Understanding what AI actually encompasses — and what distinguishes it from simpler forms of automation — is necessary for analyzing its implications in financial services.

Traditional software executes explicit instructions. A rule-based fraud detection system flags any transaction above a certain dollar threshold, or any transaction from an unfamiliar country, because a human programmer wrote those rules. The system does exactly what it was told. It cannot adapt to new fraud patterns that the rules do not address. It cannot identify complex behavioral anomalies that don't fit the prespecified rule structure. And it cannot improve based on the outcomes of its decisions.

Machine learning is the foundational AI technique that distinguishes modern AI systems from traditional rule-based software. Instead of executing explicitly programmed rules, a machine learning model learns patterns from data. It is trained on a large dataset of historical examples — past transactions labeled as fraudulent or legitimate, past loan applications labeled as defaulted or repaid — and it learns to identify the patterns that distinguish one outcome from another. The resulting model can then apply what it learned to new cases it has never seen before, generalizing from historical patterns to make predictions about new data.

Deep learning is a subset of machine learning that uses neural networks — computational architectures loosely inspired by the structure of the human brain — with many layers of interconnected nodes. Deep learning has produced the most significant AI advances of the past decade, enabling capabilities in image recognition, natural language processing, and sequential data analysis that earlier machine learning techniques could not achieve. The large language models that power conversational AI systems are deep learning models trained on enormous datasets of text, enabling them to generate human-quality language, answer questions, summarize documents, and engage in extended dialogue.

Natural language processing — NLP — is the AI capability that enables machines to understand and generate human language. In FinTech, NLP powers chatbots and virtual assistants that handle customer service interactions, document analysis systems that review contracts and regulatory filings, and sentiment analysis tools that monitor news and social media for market-moving information. The gap between early rule-based chatbots, which could only respond to specific pre-programmed inputs, and modern NLP-powered conversational AI, which can handle open-ended dialogue with considerable sophistication, illustrates how dramatically AI capabilities have advanced.

Generative AI — the category of AI that produces new content rather than simply classifying or predicting — has introduced both significant new capabilities and significant new risks in financial services. Generative AI can create realistic synthetic text, images, audio, and video. On the beneficial side, it enables sophisticated document drafting, personalized financial communications, and advanced simulation. On the harmful side, it is the technology that makes deepfakes possible — and deepfakes represent one of the most significant new fraud vectors in the FinTech sector.

AI Applications Across the FinTech Landscape

AI is deployed across virtually every function in modern FinTech, from customer acquisition through fraud prevention to regulatory compliance. The table below provides an overview of the major application areas, with examples and the key risks each creates.

Application Area	What AI Does	Representative Examples	Key Risks and Considerations
Fraud Detection	Analyzes transactions in real time to identify patterns inconsistent with legitimate behavior; adapts as fraud techniques evolve	Payment anomaly detection; account takeover identification; synthetic identity flagging; real-time authorization scoring	False positives that block legitimate transactions; adversarial fraud that deliberately mimics legitimate patterns; model drift as fraud evolves
Credit Decisioning	Incorporates alternative data sources and behavioral signals	Alternative credit scoring using transaction history, rental payments,	Algorithmic bias reproducing historical

	to assess creditworthiness beyond traditional bureau scores	and behavioral data; BNPL underwriting; small business lending decisioning	discrimination; lack of explainability for adverse actions; proxy variable problems
Customer Experience	Powers conversational interfaces, personalized product recommendations, and predictive service interventions	AI chatbots for customer service; personalized financial product recommendations; predictive alerts for overdraft risk or spending anomalies	Manipulative personalization that serves firm interests over customer interests; privacy concerns from behavioral profiling
RegTech and Compliance	Automates regulatory monitoring, reporting, and compliance processes that would require large human teams if performed manually	Automated SAR filing; real-time AML transaction monitoring; KYC document verification; regulatory change monitoring	Over-reliance on automated compliance that creates false assurance; model failures in high-stakes compliance contexts
Underwriting and Risk Pricing	Prices insurance, loan, and investment products using AI-derived risk assessments that incorporate far more variables than traditional actuarial models	AI-driven insurance pricing using telematics and behavioral data; dynamic loan pricing; algorithmic mortgage underwriting	Disparate impact in pricing; regulatory scrutiny of factors used in pricing decisions; consumer transparency concerns
Behavior Analysis	Monitors customer behavior patterns for fraud	Churn prediction models; financial stress detection;	Surveillance concerns; use of stress indicators

	prevention, product development, and retention management	engagement pattern analysis; voice and sentiment analysis in customer service	to target vulnerable customers with high-margin products
Data Security	Identifies cybersecurity threats, anomalous access patterns, and potential breaches faster than rule-based systems	AI-powered intrusion detection; behavioral biometrics for continuous authentication; deepfake detection in KYC verification	Arms race dynamic with AI-powered attacks; adversarial inputs designed to fool security models

Fraud Detection

Fraud detection is the application area where AI has delivered the most unambiguous value in financial services. The fundamental challenge of fraud detection is distinguishing fraudulent transactions from legitimate ones in real time, at scale, with low false positive rates — flagging enough fraud to prevent significant losses while avoiding blocking enough legitimate transactions to damage customer experience. Rule-based systems struggle with this balance because the rules that catch fraud also catch many legitimate transactions that happen to share surface characteristics with fraudulent ones.

Machine learning fraud detection models learn the behavioral signature of each customer — the typical amounts, locations, timing, and merchant categories of their transactions — and flag deviations from that signature for review or intervention. A transaction that is unusual for a specific customer triggers a flag even if it is entirely normal in absolute terms. This behavioral baseline approach produces substantially better fraud detection rates with lower false positive rates than rule-based alternatives, because the relevant comparison is to the individual customer's history rather than to population-wide rules.

The limitation of AI fraud detection is its dependence on the quality and representativeness of its training data. A model trained primarily on fraud patterns from one demographic group or transaction type may perform poorly when confronted with fraud targeting different groups or using different techniques. Sophisticated fraudsters study fraud detection systems and deliberately craft transactions designed to mimic legitimate behavior — a practice called adversarial attack — which requires continuous model retraining and monitoring to address.

Credit Decisioning

AI credit decisioning expands the information available for credit assessment beyond the traditional inputs of credit bureau scores, income, and employment — incorporating transaction history, rental payment records, utility payments, behavioral signals, and in some cases thousands of additional variables. The promise is that alternative data sources can identify creditworthy borrowers who are invisible to traditional models because they have thin credit files — recent immigrants, young adults, people recovering from past financial difficulties — expanding credit access in ways that serve both borrowers and lenders.

The risk is that alternative data sources may encode historical discrimination in ways that are not immediately visible. A credit model that uses zip code as a variable is incorporating a proxy for race, because residential segregation has produced zip codes that are highly correlated with racial composition. A model that uses educational attainment is incorporating a proxy for socioeconomic background. A model that uses the device type used to apply — whether the applicant used an iPhone or a lower-cost Android device — is potentially incorporating proxies for income and race that the model was not explicitly designed to use. The Equal Credit Opportunity Act prohibits discrimination in credit on the basis of race, color, religion, national origin, sex, marital status, and age — but it applies to the outcomes of credit decisions, not to the specific variables that a model uses, which means that discriminatory outcomes can result from models that use no explicitly prohibited variables.

Customer Experience and Conversational AI

AI-powered customer service — chatbots, virtual assistants, and automated service resolution systems — has become standard infrastructure in FinTech. The economic logic is straightforward: a well-designed AI system can handle the majority of routine customer inquiries — balance checks, transaction disputes, password resets, product information — at a fraction of the cost of human agents, at any hour of the day, with unlimited scalability. Human agents are then available for complex inquiries that require judgment, empathy, or authority to make exceptions.

The customer experience implications of AI service are more complicated than the cost story suggests. An AI system that cannot escalate appropriately to a human when it encounters a situation outside its competence — or that makes escalation deliberately difficult to reduce cost — creates customer frustration and potential harm. A personalization system that recommends financial products based on a model designed to maximize the firm's revenue rather than the customer's financial outcomes is using AI against the customer's interests. The ethical framework from Chapter 7 applies directly: personalization that exploits behavioral data to target vulnerable customers with high-margin products treats those customers as means rather than ends.

Algorithmic Bias: How It Enters, Why It Persists

Algorithmic bias is the systematic tendency of an AI model to produce outputs that are unfair or discriminatory with respect to identifiable groups. It is not random error — it is a consistent pattern in which the model performs differently for different groups, typically in ways that disadvantage groups that have historically faced discrimination. Understanding how bias enters AI systems and why it persists despite awareness of the problem is essential for anyone responsible for building or deploying AI in financial services.

Sources of Bias

The most common source of algorithmic bias is biased training data. A machine learning model learns patterns from historical data. If that historical data reflects historical discrimination — lending decisions made by human loan officers who denied credit to qualified minority applicants, hiring decisions that systematically favored men, medical diagnoses made by systems that underdiagnosed conditions in women or people of color — the model will learn those discriminatory patterns and reproduce them. The model is not discriminating intentionally. It is doing exactly what it was trained to do: identify the patterns in the data. If the patterns in the data are discriminatory, the model's outputs will be discriminatory.

A well-documented example involves voice recognition systems. Early voice recognition technology was developed primarily by teams of male engineers who tested their systems primarily on male voices. The resulting systems performed significantly better on male voices than on female voices — not because of any intentional design decision, but because the training data and the testing population were both skewed toward men. A voice-recognition system that fails to recognize female speakers most likely does so because the program was primarily developed and tested by men. The lesson generalizes: any AI system trained and tested on a population that does not reflect the diversity of the population it will serve is likely to perform worse for the underrepresented groups.

Proxy variables are a second major source of bias. A proxy variable is a variable that is correlated with a protected characteristic — race, gender, age, national origin — without explicitly measuring it. Using zip code as a predictor in a credit model is using a proxy for race. Using educational institution attended is using a proxy for socioeconomic background. The model never sees the protected characteristic directly; it learns to use the correlated proxy instead, producing outcomes that are functionally equivalent to explicit discrimination even though no explicitly prohibited variable appears in the model.

Feedback loops are a third source of bias that is particularly difficult to address. A credit model that denies credit to people in certain zip codes produces fewer credit histories from those zip codes, which means the next version of the model has less data about creditworthiness in those areas, which means its predictions for those areas are less accurate and more likely to continue denying credit. The model's outputs influence the data that is used to train the next version of the model, and discriminatory patterns can become self-reinforcing over time even when each individual model update is technically sound.

Why Bias Is Hard to Detect and Correct

Algorithmic bias is difficult to detect for several reasons. Complex machine learning models — particularly deep learning models with millions of parameters — do not produce simple, interpretable decision rules. There is no single coefficient to examine that reveals whether the model is treating groups differently. Detecting bias requires deliberate testing: running the model on representative test populations, disaggregating its outputs by demographic group, and examining whether performance metrics differ significantly across groups. This testing must be designed and executed explicitly — it does not happen automatically.

The software development process itself can reproduce bias if it does not deliberately counteract it. Software designers should include users in the design stages — a principle that is straightforward to state and systematically neglected in practice. AI systems are built by teams whose demographic composition often does not reflect the diversity of the populations the systems will affect. An engineering team that is overwhelmingly male and from a particular socioeconomic background may not identify bias that would be obvious to a more diverse team, because the affected users are not represented in the development process. Diverse development teams, user testing that includes representative populations, and explicit bias auditing at each stage of development are the structural responses to this problem — but they require deliberate investment that many development processes do not make.

Correcting bias once it is identified creates its own complications. Simply removing a biased variable from the model does not eliminate the bias if correlated proxy variables remain. Reweighting training data to reduce historical discrimination can improve fairness metrics on one dimension while degrading them on others. Fairness itself is not a single concept — different mathematical definitions of fairness are sometimes mutually inconsistent, meaning that a model that satisfies one definition of fairness may necessarily fail another. These technical complexities do not excuse inaction, but they do explain why bias elimination is not as straightforward as identifying the problem and fixing it.

The Apple Card Case: Bias in Credit Decisioning

In November 2019, David Heinemeier Hansson, a software developer and entrepreneur, posted on social media that Apple Card had given him a credit limit twenty times higher than the limit given to his wife, despite the fact that they filed joint tax returns and his wife had a higher personal credit score. The post went viral. Other users reported similar experiences — women receiving lower credit limits than their male partners despite equal or superior credit profiles. The New York Department of Financial Services opened an investigation.

Apple Card was issued by Goldman Sachs, which was responsible for the credit decisioning algorithm. Goldman Sachs responded that the algorithm did not use gender as a variable — a response that was technically accurate but that missed the core concern. The ECOA does not prohibit only the explicit use of prohibited variables. It prohibits outcomes that constitute discrimination on prohibited bases. An algorithm that produces systematically lower credit limits for women than for comparably situated men is producing discriminatory outcomes regardless of whether the word gender appears in the model's variable list.

The investigation by the New York DFS ultimately found no evidence that Goldman Sachs's algorithm violated fair lending laws — a conclusion that itself illuminates the limitations of existing regulatory frameworks. The investigation focused on whether the algorithm used prohibited variables and whether it produced outcomes that could be explained by legitimate credit factors. The agency did not have clear authority to require the kind of granular fairness testing — disaggregating model performance by gender across comparable credit profiles — that would have provided the most direct evidence of whether the algorithm was producing discriminatory outcomes.

The Apple Card case illustrates several important points. First, the absence of explicitly prohibited variables does not guarantee the absence of discriminatory outcomes. Second, the married couple testing methodology — comparing credit decisions for spouses — is an imperfect proxy for systematic bias because it conflates individual decisions with systemic patterns. Third, the regulatory framework that exists for traditional credit discrimination is imperfectly suited to AI-driven credit decisioning, because it was designed to address human decisions and does not straightforwardly map onto the opaque multi-variable models that AI systems use. Fourth, and perhaps most importantly, the reputational consequences of an algorithmic bias allegation — even one that does not ultimately result in a regulatory finding — can be severe and immediate in ways that create strong incentives for proactive fairness testing.

Deepfakes and AI-Enabled Fraud

A deepfake is a synthetic media product — video, audio, or image — generated by AI to make it appear that a real person said or did something they did not say or do. The term derives from deep learning, the AI technique used to generate the synthetic content, and fake, describing the fabricated nature of the output. Deepfake technology has advanced rapidly, and the quality of synthetic media has reached the point where trained observers can struggle to distinguish AI-generated content from authentic recordings.

In the FinTech context, deepfakes represent a significant and growing fraud vector with several distinct applications. Voice cloning — generating a synthetic audio recording that sounds like a specific person — has been used to deceive financial institutions in social engineering attacks. In a documented case, criminals cloned the voice of a company executive and used it in a phone call to a finance employee, instructing the employee to transfer funds to an external account. The employee, believing they were speaking with their CEO, complied. The fraud succeeded because the synthetic voice was convincing enough that the human element of the authentication process — listening to the caller — failed.

Synthetic identity fraud using AI-generated faces has become a significant concern for KYC processes that rely on identity document verification and facial matching. An AI-generated photograph can create a plausible human face that does not belong to any real person, potentially usable to construct a synthetic identity that passes automated document verification. Liveness detection — technology designed to confirm that the person submitting a verification selfie is physically present rather than submitting a photograph — has become an AI arms race, as deepfake technology capable of generating convincing video has challenged the reliability of liveness checks designed for simpler forms of spoofing.

Deepfake phishing — using synthetic audio or video of trusted individuals to make phishing communications more convincing — extends the social engineering threat in ways that traditional security awareness training did not anticipate. A video call in which a person appears to see and hear a trusted colleague or executive asking them to take an action creates a far more compelling impersonation than an email claiming to be from that person. As video deepfake technology becomes more accessible, the assumption that a video call with a familiar face is authentic becomes increasingly unreliable.

The regulatory and technical response to deepfake fraud is developing rapidly. Watermarking and provenance standards — technical mechanisms for authenticating the origin of digital media — are being developed by technology companies and standards bodies. The Coalition for Content Provenance and Authenticity, or C2PA, has developed open standards for attaching tamper-evident metadata to digital content that allows recipients to verify its origin and whether it has been modified. In financial services, the practical response has focused on multi-factor authentication that does not rely on voice or face recognition alone, on callback verification procedures that confirm identity through independent channels, and on employee training specifically addressing deepfake fraud techniques.

RegTech: AI in the Service of Compliance

RegTech — regulatory technology — refers to the application of technology, and increasingly AI, to regulatory compliance functions. The compliance burden in financial services is substantial and growing: the volume of regulatory requirements, the complexity of the reporting obligations, and the pace at which regulations change all create compliance costs that are significant even for large institutions and potentially prohibitive for smaller FinTech firms. AI applied to compliance automation offers the prospect of meeting these obligations more efficiently and more accurately than manual processes permit.

AML transaction monitoring is the RegTech application with the longest history and the most significant AI penetration. Traditional rule-based AML monitoring generates enormous volumes of false positives — estimates suggest that ninety to ninety-five percent of AML alerts generated by rule-based systems are false positives that require analyst review but do not result in SAR filings. Machine learning monitoring systems, trained on historical data about both legitimate and suspicious transactions, can substantially reduce false positive rates while maintaining or improving detection rates. The business case is strong: reducing false positives frees compliance staff to focus on genuine suspicious activity rather than processing alerts that will be cleared immediately.

KYC verification automation uses AI-powered document scanning, optical character recognition, and facial matching to accelerate identity verification at onboarding. A process that once required manual review of identity documents — and might take days in a traditional bank — can be completed in minutes with automated verification. AI can also screen customer names against OFAC sanctions lists, watch lists, and adverse media databases at a speed and scale that manual review cannot match. The same technology that makes frictionless digital onboarding possible is also what makes it possible for firms to meet KYC obligations at the volume that digital financial services require.

Regulatory change monitoring — AI systems that track regulatory publications, guidance, and enforcement actions and alert compliance teams to changes relevant to their specific business — addresses the challenge of keeping pace with a regulatory environment that evolves continuously. A payment platform operating across multiple states and jurisdictions faces different regulatory requirements in each, and those requirements change as legislatures act, agencies issue guidance, and courts rule. AI systems that monitor this landscape and flag relevant changes reduce the risk that regulatory evolution goes unnoticed until a compliance failure occurs.

The irony of RegTech is that the same AI technology raising the most significant regulatory concerns — for bias, for explainability, for accountability — is also the most powerful tool available for meeting the regulatory obligations those concerns are generating. Firms that invest in AI-powered compliance infrastructure are simultaneously the firms best positioned to comply with the AI-specific regulations that are emerging. The relationship between AI risk and AI compliance capability is not coincidental — it reflects the reality that complex AI systems require sophisticated AI tools to monitor and govern effectively.

The Black Box Problem: Explainability and Accountability

A credit officer who denies a loan application can, when required, explain why: the applicant's debt-to-income ratio exceeded policy limits, or the credit score fell below the minimum threshold, or the employment history was insufficient. The explanation may not satisfy the applicant, but it is articulable. A deep learning credit model that denies the same application based on patterns identified across hundreds of variables may produce an output — deny — without producing an explanation that is either meaningful to the applicant or auditable by a regulator.

This is the black box problem: complex AI models produce accurate predictions while generating limited insight into why they produce those predictions. The opacity is not an accident or a design flaw — it is inherent to how the most powerful machine learning techniques work. A deep neural network that identifies fraud by detecting subtle patterns across thousands of simultaneous variables is valuable precisely because it can detect those patterns. The same complexity that makes the detection possible makes the explanation difficult.

The Legal Right to Explanation

The Equal Credit Opportunity Act requires creditors who take adverse action on a credit application to provide the applicant with a statement of reasons — the specific factors that caused the adverse action. This requirement was designed for human credit decisions, where the loan officer could identify the key factors. Applied to AI credit models, the adverse action notice requirement creates a legal obligation for explanation that complex models may not be able to satisfy straightforwardly.

Regulatory guidance from the CFPB and federal banking regulators has addressed this tension, requiring that AI-driven credit decisions be capable of generating compliant adverse action notices — specific, accurate statements of the factors that led to the adverse outcome. This requirement has practical design implications: credit models must be built or supplemented with explainability tools capable of identifying the variables that most influenced each individual decision. A model that cannot generate this explanation cannot be used for consumer credit decisions under current U.S. law, regardless of its predictive accuracy.

The EU's General Data Protection Regulation, addressed in Chapter 8, includes a right to explanation for automated decisions that significantly affect individuals. Under GDPR, individuals subject to fully automated decisions have the right to request a meaningful explanation of the logic involved. This requirement has shaped AI development practices for firms operating in or serving European markets, and it represents the most developed legal framework for algorithmic accountability currently in effect internationally.

Model Governance

Model governance is the set of policies, processes, and controls that a firm uses to develop, validate, deploy, monitor, and retire AI models. It is the internal governance framework for AI systems, analogous to the broader corporate governance framework covered in Chapter 12 but specific to the AI systems that increasingly drive consequential decisions in FinTech.

Effective model governance begins before a model is built, with requirements that define what the model is intended to do, what data it will use, what fairness constraints it must satisfy, and how it will be tested before deployment. Model validation — independent review of a model's methodology, data, assumptions, and outputs before deployment — is the key control that separates rigorous model governance from model development without oversight. A model validated only by the team that built it has not received independent scrutiny. A model reviewed by an independent validation function that has the authority to require changes before deployment has.

Post-deployment monitoring tracks model performance over time, identifying when a model's accuracy or fairness characteristics have degraded — a phenomenon called model drift — and triggering retraining or replacement when performance falls below acceptable thresholds. A fraud detection model trained on pre-pandemic transaction patterns may perform poorly when consumer behavior changes significantly. A credit model trained during a period of low default rates may underestimate credit risk when economic conditions deteriorate. Continuous monitoring is the mechanism through which these degradations are detected before they cause significant harm.

The Regulatory Landscape for AI in Financial Services

The regulatory framework governing AI in financial services is fragmented, evolving, and — particularly in the United States — substantially less developed than the AI capabilities it is trying to address. Understanding what exists, what is emerging, and what gaps remain is essential for FinTech professionals designing AI systems that will need to operate in this environment.

The U.S. Approach: Existing Laws Applied to New Technology

The United States does not have a comprehensive federal AI law applicable to financial services as of 2025. Instead, existing laws are being applied to AI-driven decisions through regulatory guidance, enforcement actions, and reinterpretation of existing statutory authority. The ECOA and its implementing regulation, Regulation B, prohibit discriminatory outcomes in credit decisioning and require adverse action notices — and these requirements apply to AI-driven credit decisions regardless of how the model works. The FTC Act's prohibition on unfair or deceptive acts or practices applies to AI-driven marketing, pricing, and customer service. The Fair Housing Act prohibits discriminatory outcomes in mortgage lending that apply equally to algorithmic underwriting.

The CFPB has issued guidance on the use of AI in credit decisioning, affirming that ECOA's requirements apply fully to algorithmic models and clarifying that adverse action notices must identify the specific factors that most significantly contributed to an adverse outcome — not generic statements about the model's overall operation. The CFPB's 2023 circular on algorithmic models explicitly rejected the argument that complex AI models are exempt from adverse action notice requirements because they cannot generate specific factor explanations.

The federal banking regulators — the OCC, FDIC, Federal Reserve, and NCUA — issued interagency guidance on model risk management in 2011, updated for AI in subsequent guidance, that establishes expectations for model validation and governance at supervised institutions. While this guidance predates the current generation of AI, its core principles — independent validation, ongoing monitoring, clear governance of model development and deployment — apply directly to AI systems and are being applied in bank examinations.

The Federal Trade Commission has signaled active interest in AI governance through enforcement actions and policy statements. The FTC's AI guidelines emphasize that existing FTC authority — covering deceptive practices, unfair competition, and consumer harm — applies to AI systems, and that algorithmic systems that produce discriminatory or deceptive outcomes are subject to FTC enforcement regardless of the underlying technology. The EEOC has issued guidance on AI in hiring and employment decisions, noting that employment AI tools that produce disparate impact may violate Title VII of the Civil Rights Act.

The EU AI Act: An International Benchmark

The European Union's AI Act, adopted in 2024 and entering into force in stages through 2027, is the most comprehensive AI regulatory framework in the world and represents the international benchmark against which U.S. regulatory development is increasingly measured. The AI Act takes a risk-based approach: AI systems are classified by the risk level of their applications, with higher-risk applications subject to more stringent requirements.

AI systems used in credit scoring, insurance pricing, and other financial services decisions are classified as high-risk under the AI Act, triggering obligations that include: mandatory conformity assessment before deployment, transparency requirements for affected individuals, human oversight requirements that ensure a human can review and override AI decisions, accuracy and robustness standards, and registration in an EU database of high-risk AI systems. These requirements will apply to U.S. FinTech firms that offer services to EU customers — and they are influencing AI development practices globally, because firms serving both U.S. and EU markets typically build to the higher standard.

The AI Act also prohibits certain AI applications entirely — including real-time biometric identification in public spaces for law enforcement, AI systems that exploit psychological vulnerabilities to manipulate behavior, and social scoring systems that evaluate individuals based on their behavior. While these prohibitions are not directly applicable to most FinTech applications, they establish the EU's position that some AI applications are unacceptable regardless of their commercial value.

State-Level AI Regulation

In the absence of comprehensive federal AI legislation, states have begun enacting their own AI requirements, creating the same kind of patchwork regulatory environment that characterizes U.S. data privacy law. Colorado's AI Act, effective in 2026, requires developers and deployers of high-risk AI systems to use reasonable care to protect consumers from algorithmic discrimination, conduct impact assessments, and provide disclosures to affected individuals. Several other states have enacted or are considering similar requirements. For FinTech firms operating nationally, the emergence of state AI regulation creates compliance obligations that vary by jurisdiction and that, in the absence of federal preemption, will continue to diverge.

Building Responsible AI: Firm Obligations and Professional Accountability

The regulatory landscape for AI is incomplete and evolving, but the obligations of FinTech firms in deploying AI systems are not simply whatever the law currently requires. Chapter 7's discussion of the ACM/IEEE Software Engineering Code of Ethics applies directly here: engineers and data scientists who build AI systems bear professional responsibility for those systems' outputs and their effects on the people those outputs affect. The gap between what is currently required and what professional ethics demands is often significant.

Fairness Testing and Bias Auditing

A FinTech firm that deploys a credit model without testing it for disparate impact across demographic groups has not met its professional obligations, regardless of whether current law explicitly requires such testing. Fairness testing involves running the model on representative test populations, disaggregating its outputs by protected characteristics, and examining whether error rates, approval rates, and pricing outcomes differ significantly across groups. Where they do, the firm has an obligation to investigate whether the disparity reflects legitimate credit factors or proxy variable bias, and to take corrective action where the latter is the case.

Third-party bias audits — independent assessments of an AI model's fairness characteristics by parties outside the firm — provide a form of accountability that internal testing alone cannot. Just as financial audits provide independent verification of financial statements that internal accounting cannot, external AI audits provide independent scrutiny of model fairness claims. Several firms now offer AI auditing services specifically for financial services models, and regulatory guidance in some jurisdictions is beginning to require or encourage independent audit as a component of AI governance.

Human Oversight and the Limits of Automation

The efficiency case for AI — faster decisions, lower costs, unlimited scalability — can create pressure to automate decisions that should retain meaningful human oversight. A credit decision that affects an applicant's ability to buy a home, fund a business, or manage a financial emergency is consequential enough that the person affected by it should have access to human review of an AI-generated outcome they believe is wrong.

Effective human oversight is not the same as nominal human oversight. A review process in which a human approves AI recommendations without the time, information, or authority to meaningfully question them is not providing real oversight — it is providing a human signature on an AI decision. Real oversight requires that reviewers have access to the information needed to evaluate the recommendation, authority to override it, time to exercise genuine judgment, and incentives aligned with making the right decision rather than processing volume quickly. Designing human oversight that actually functions — rather than simply satisfying a regulatory checkbox — is a governance design challenge as much as a technical one.

Transparency and Consumer Communication

Consumers who are subject to AI-driven decisions have a legitimate interest in understanding that AI was used, what factors influenced the outcome, and how they can seek recourse. The adverse action notice requirement provides a legal floor for this transparency in the credit context, but professional practice should extend beyond the legal floor to genuine transparency about how AI is used in customer-facing decisions.

This transparency obligation has a design dimension. AI systems that are built to be explainable from the outset — using techniques such as model-agnostic explanation methods, simpler interpretable models where accuracy requirements permit, and decision support architectures that keep a human in the loop for high-stakes decisions — are better positioned to meet transparency obligations than systems built purely for predictive accuracy without explainability as a design constraint. The choice to prioritize explainability in design is a professional and ethical choice, not merely a compliance response.

Key Takeaways

Artificial intelligence in FinTech encompasses a range of techniques — machine learning, deep learning, natural language processing, and generative AI — that differ significantly from traditional rule-based software. Where traditional software executes explicit instructions, machine learning models learn patterns from data and generalize those patterns to new cases. This distinction matters because the capabilities, limitations, and risks of AI systems are fundamentally different from those of the rule-based systems they are increasingly replacing.

The major applications of AI in FinTech include fraud detection, credit decisioning, customer experience and conversational AI, RegTech and compliance automation, underwriting and risk pricing, behavior analysis, and data security. Each application area combines genuine capability — faster, more accurate, more scalable than manual alternatives — with distinctive risks including bias, opacity, adversarial vulnerability, and the potential for automation to amplify harm at a scale that human processes cannot match.

Algorithmic bias is the systematic tendency of an AI model to produce outputs that are unfair or discriminatory with respect to identifiable groups. It enters models through biased training data that reflects historical discrimination, through proxy variables correlated with protected characteristics, and through feedback loops that reinforce discriminatory patterns over time. A voice recognition system that fails to recognize female speakers most likely does so because the program was primarily developed and tested by men — a representative example of how underrepresentation in the development process produces systems that perform poorly for underrepresented groups. Software designers should include users in the design stages; diverse development teams, representative testing populations, and explicit bias auditing at each stage of development are the structural responses to this problem.

The Apple Card case — in which Goldman Sachs's credit algorithm was alleged to have given women systematically lower credit limits than comparably situated men — illustrates the limits of both the defense that no prohibited variable was used and of existing regulatory frameworks designed for human credit decisions. The case established that algorithmic bias allegations create immediate reputational consequences and regulatory scrutiny even when they do not ultimately result in enforcement findings, creating strong incentives for proactive fairness testing.

Deepfakes — synthetic media generated by AI to impersonate real people — represent a significant and growing fraud vector in FinTech. Voice cloning has been used to impersonate executives in social engineering attacks that resulted in unauthorized fund transfers. AI-generated synthetic identities challenge KYC verification systems. Deepfake video challenges the assumption that a video call with a familiar face is authentic. The technical response includes watermarking and content provenance standards; the operational response requires multi-factor authentication that does not rely on voice or face recognition alone.

RegTech — the application of AI to regulatory compliance — offers the prospect of meeting complex compliance obligations more efficiently and accurately than manual processes. AI-powered AML transaction monitoring substantially reduces false positive rates relative to rule-based systems. KYC verification automation enables compliant identity verification at the scale digital onboarding requires. Regulatory change monitoring reduces the risk that evolving requirements go unnoticed. The same AI technology raising governance concerns is also the most powerful tool available for meeting the compliance obligations those concerns are generating.

The black box problem — the opacity of complex AI models that produce accurate predictions without generating interpretable explanations — creates legal and ethical obligations for FinTech firms. The Equal Credit Opportunity Act requires adverse action notices with specific factor explanations that apply to AI credit models. The CFPB has affirmed that this requirement cannot be satisfied by generic statements about model operation. The EU AI Act, the most comprehensive international AI regulatory framework, classifies credit scoring as a high-risk AI application requiring conformity assessment, transparency disclosures, and meaningful human oversight before deployment.

The responsibilities of FinTech firms in deploying AI extend beyond current legal requirements to the professional ethics obligations that engineers and data scientists bear for the systems they build. These responsibilities include proactive fairness testing and bias auditing, model governance that includes independent validation and continuous post-deployment monitoring, human oversight mechanisms that provide genuine rather than nominal review of high-stakes AI decisions, and transparency with consumers about how AI is used in decisions that affect them. Building AI systems that are accurate, fair, explainable, and accountable is not merely a compliance project — it is the professional standard to which practitioners in this field are held.

Glossary

The following definitions reflect the usage of key terms throughout this textbook. Terms are defined in the context of FinTech, data ethics, and U.S. regulatory practice. For terms that have broader meanings outside this context, the definitions here are specific to how the term is used in this course.

A

Adverse Action Notice — A written notice required by the Equal Credit Opportunity Act and the Fair Credit Reporting Act that informs a consumer when credit has been denied, reduced, or terminated, and identifies the specific reasons for the decision. Required for AI-driven credit decisions as well as traditional underwriting.

Alert Fatigue — A condition in which compliance personnel process automated alerts quickly and without adequate scrutiny because alert volume exceeds their capacity for meaningful review, resulting in genuine suspicious activity being missed.

Algorithmic Bias — Systematic and unfair discrimination produced by an algorithm, typically resulting from biased training data, flawed model design, or proxy variables that correlate with protected characteristics such as race or gender. A central ethical and legal concern in AI-driven financial services.

AML (Anti-Money Laundering) — The body of laws, regulations, and procedures designed to prevent criminals from disguising illegally obtained funds as legitimate income. In the United States, AML requirements are primarily governed by the Bank Secrecy Act and administered by FinCEN.

API (Application Programming Interface) — A set of protocols and tools that allows different software applications to communicate with each other. In FinTech, APIs enable third-party apps to access financial data and services with user permission, and are central to open banking frameworks.

Automated Clearing House (ACH) – The electronic network used for financial transactions in the United States, including direct deposits, payroll, bill payments, and business-to-business payments. ACH transactions are processed in batches and typically settle within one to two business days.

B

Bank Secrecy Act (BSA) – The foundational U.S. anti-money laundering statute, enacted in 1970, that requires financial institutions to maintain records and file reports to help law enforcement detect and prosecute money laundering and financial crime. Administered by FinCEN.

BitLicense – A regulatory license for virtual currency businesses operating in New York, introduced by the New York Department of Financial Services in 2015. The first state-level regulatory framework specifically designed for cryptocurrency businesses.

Black Box Problem – The challenge of explaining how a complex AI model arrived at a particular decision when the model's internal logic is too complex to interpret directly. A significant issue for AI systems used in high-stakes decisions such as credit underwriting.

Blockchain – A distributed ledger technology in which transactions are recorded in linked blocks that are cryptographically secured and replicated across a network of computers. Once recorded, transactions cannot be altered without changing all subsequent blocks, making the ledger tamper-resistant.

Broker-Dealer – A firm or individual that buys and sells securities either as an agent on behalf of customers or as a principal for its own account. Broker-dealers are regulated by the SEC and FINRA and are subject to Regulation Best Interest when making recommendations to retail customers.

Buy Now Pay Later (BNPL) — A short-term financing product that allows consumers to make purchases and pay for them in installments, typically with no interest if paid within a specified period. BNPL products have attracted regulatory attention from the CFPB regarding their consumer protection implications.

C

CFPB (Consumer Financial Protection Bureau) — A federal agency created by the Dodd-Frank Act of 2010 with the mandate to protect consumers in the market for financial products and services. Unlike most federal financial regulators, the CFPB has direct authority over nonbank FinTech firms that offer consumer financial products.

CFTC (Commodity Futures Trading Commission) — The federal agency that regulates U.S. derivatives markets, including futures, options, and swaps. The CFTC has taken the position that Bitcoin and Ethereum are commodities and has jurisdiction over cryptocurrency derivatives products.

CISA (Cybersecurity Information Sharing Act) — A federal law enacted in 2015 that authorizes and encourages private entities to share cyber threat indicators and defensive measures with each other and with the federal government, with legal protections for entities that share in accordance with the statute.

Consent — In data privacy, the voluntary, informed, and specific agreement by an individual to the collection, use, or sharing of their personal data. A foundational concept in both ethical data practice and legal frameworks such as the GDPR.

Corrective Controls — Internal controls designed to identify and remedy problems after they have occurred, and to restore normal operations. Examples include incident response plans, data recovery procedures, and disciplinary processes for policy violations.

Cryptocurrency — A digital or virtual currency that uses cryptography for security and operates on a decentralized network, typically a blockchain. Cryptocurrencies are not issued or controlled by any central authority. Bitcoin and Ethereum are the most widely known examples.

Customer Due Diligence (CDD) — The process by which financial institutions collect and verify information about customers to understand the nature and purpose of their relationships and assess the risk that those relationships may be used for money laundering or financial crime.

Customer Identification Program (CIP) — The minimum KYC obligation under the Bank Secrecy Act, requiring financial institutions to collect and verify identifying information from customers before opening an account, including name, date of birth, address, and identification number.

Cyber Threat Indicator — Under CISA, information that describes or identifies a malicious reconnaissance activity, security vulnerability, method of defeating a security control, or other attribute of a cybersecurity threat. Sharing cyber threat indicators is the primary mechanism through which CISA encourages collective defense.

D

DAO (Decentralized Autonomous Organization) — An organization governed by rules encoded in smart contracts on a blockchain, with decisions made through token holder voting rather than traditional management structures. The SEC's 2017 DAO Report concluded that DAO tokens can constitute securities under the Howey Test.

Data Ethics — The branch of ethics concerned with the appropriate collection, use, sharing, and protection of data, particularly personal data. In FinTech, data ethics addresses questions about consent, purpose limitation, transparency, and the balance between commercial data use and individual privacy rights.

Data Minimization — The principle that organizations should collect only the personal data that is necessary for the specific purpose for which it is collected, and should not retain it longer than necessary. A core principle of the GDPR and a foundational concept in ethical data practice.

Decentralized Exchange (DEX) — A cryptocurrency exchange that operates without a central intermediary, using smart contracts to facilitate trades directly between users. DEXs present significant regulatory challenges because they may not have an identifiable operator subject to KYC and AML requirements.

Deepfake — Synthetic media — typically video or audio — generated by artificial intelligence to create realistic but fabricated representations of real people. In FinTech, deepfakes are used in fraud schemes including identity verification bypass and CEO impersonation.

Deontological Ethics — An ethical framework that evaluates the morality of actions based on whether they conform to rules or duties, regardless of their consequences. Associated with philosopher Immanuel Kant and the categorical imperative: act only according to rules you would want universally applied.

Detective Controls — Internal controls designed to identify problems, errors, or irregularities after they have occurred, enabling timely response. Examples include transaction monitoring systems, reconciliation processes, internal audits, and security logging.

Distributed Ledger Technology (DLT) — A category of technology in which records are stored and synchronized across multiple locations without a central administrator. Blockchain is the most widely known form of distributed ledger technology.

Dodd-Frank Act — The Dodd-Frank Wall Street Reform and Consumer Protection Act of 2010, enacted in response to the 2008 financial crisis. Dodd-Frank created the CFPB, expanded financial regulatory oversight, and introduced significant consumer protection and systemic risk provisions.

Duty of Care — One of the two components of fiduciary duty, requiring an investment adviser or other fiduciary to provide advice that is suitable and well-reasoned given what the adviser knows about the client's goals, risk tolerance, and financial situation.

Duty of Loyalty — One of the two components of fiduciary duty, requiring an investment adviser or other fiduciary to avoid conflicts of interest or, where conflicts cannot be avoided, to disclose and manage them so that the client's interests are not subordinated.

E

ECOA (Equal Credit Opportunity Act) — A federal law that prohibits creditors from discriminating against credit applicants on the basis of race, color, religion, national origin, sex, marital status, age, or receipt of public assistance. ECOA's requirements apply to AI-driven credit decisions as well as traditional underwriting.

Enhanced Due Diligence (EDD) — A heightened level of KYC scrutiny applied to higher-risk customer relationships, including politically exposed persons, customers in high-risk jurisdictions, and customers whose activity is inconsistent with their stated purpose.

ERP System (Enterprise Resource Planning) — Integrated software that manages core business processes across an organization. In financial services, ERP systems merge high-value data with operational processes, making them targets for cyberattacks and requiring robust access controls and audit trails.

F

FCRA (Fair Credit Reporting Act) — A federal law that regulates how consumer credit information is collected, used, and shared. The FCRA requires consumer reporting agencies to ensure the accuracy of credit information, limits the permissible purposes for accessing credit reports, and requires adverse action notices when credit is denied.

FDIC (Federal Deposit Insurance Corporation) — The federal agency that insures deposits at U.S. banks and savings institutions up to $250,000 per depositor. The FDIC also supervises state-chartered banks that are not members of the Federal Reserve System and has approval authority over industrial loan company applications.

FedNow — The Federal Reserve's real-time payment and settlement service, launched in 2023, that enables financial institutions to offer customers the ability to send and receive payments instantly at any time, including weekends and holidays.

Fiat Currency — Government-issued currency that is not backed by a physical commodity such as gold, but derives its value from government declaration and public trust. The U.S. dollar, euro, and Japanese yen are examples of fiat currencies.

Fiduciary Duty — A legal obligation to act in the best interest of another party, placing their interests ahead of your own. In investment management, fiduciary duty applies to registered investment advisers and has two components: the duty of loyalty and the duty of care.

FinCEN (Financial Crimes Enforcement Network) — A bureau of the U.S. Department of the Treasury that administers the Bank Secrecy Act, collects and analyzes financial intelligence, and combats money laundering and terrorism financing. FinCEN classifies virtual currency exchangers as money services businesses subject to BSA requirements.

FinTech — Short for financial technology. The application of technology to improve, automate, or disrupt the delivery of financial services. FinTech encompasses payments, lending, cryptocurrency, investment platforms, insurance technology, and regulatory technology, among other sectors.

Firewall — A network security system that monitors and controls incoming and outgoing network traffic based on predetermined security rules. Firewalls are a foundational cybersecurity control that establish a barrier between trusted internal networks and untrusted external networks.

Form ADV — The disclosure document that registered investment advisers must file with the SEC, describing their business, fees, conflicts of interest, disciplinary history, and the qualifications of key personnel. Form ADV is publicly available through the SEC's Investment Adviser Public Disclosure database.

FTC (Federal Trade Commission) — The federal agency that enforces consumer protection laws prohibiting unfair or deceptive acts or practices in commerce. The FTC has broad authority over FinTech companies' data practices, privacy representations, and consumer-facing conduct regardless of their industry.

FTX — A centralized cryptocurrency exchange that collapsed in November 2022 after it was revealed that customer funds had been misappropriated to cover losses at its affiliated trading firm, Alameda Research. The collapse resulted in the criminal conviction of founder Sam Bankman-Fried and became a defining regulatory event for the cryptocurrency industry.

G

GDPR (General Data Protection Regulation) — The European Union's comprehensive data protection law, effective since 2018, that applies to any organization collecting or processing the personal data of EU residents. The GDPR establishes broad individual rights including access, portability, erasure, and the right to explanation for automated decisions.

GLB Act (Gramm-Leach-Bliley Act) — A federal law requiring financial institutions to explain their information-sharing practices to customers and to safeguard sensitive data. The GLB Act includes a Privacy Rule governing how customer financial information is shared and a Safeguards Rule governing how it is protected.

Governance — The systems, structures, and processes by which an organization is directed, controlled, and held accountable. In FinTech, corporate governance encompasses board oversight, compliance functions, risk management, and the internal controls that ensure the organization operates within legal and ethical boundaries.

H

Hacktivism — The use of hacking techniques to advance a political or social agenda. Hacktivists typically target organizations they consider to be acting against the public interest, using tactics such as website defacement and distributed denial of service attacks.

Hash Function — A cryptographic function that converts input data of any size into a fixed-size output, called a hash or digest. In blockchain, hash functions link blocks together and make tampering detectable by ensuring that any change to a block changes its hash.

Honey Pot — A cybersecurity tool that mimics a vulnerable system or data store to attract and detect attackers, gather intelligence about attack methods, and divert attackers from real systems. Honey pots are a detective control used to identify intrusions and study attacker behavior.

Howey Test — The legal standard drawn from the 1946 Supreme Court case SEC v. W.J. Howey Co. for determining whether a financial arrangement constitutes an investment contract and therefore a security under federal law. An investment contract exists when there is an investment of money in a common enterprise with an expectation of profit derived from the efforts of others.

I

ICO (Initial Coin Offering) — A fundraising mechanism in which a company or project issues digital tokens to investors, typically in exchange for established cryptocurrencies such as Bitcoin or Ethereum. ICOs have been subject to significant SEC scrutiny on the grounds that many token offerings constitute unregistered securities offerings.

ILC (Industrial Loan Company) — A state-chartered financial institution that can accept federally insured deposits but is not subject to Federal Reserve oversight as a bank holding company, allowing commercial companies to own banking capabilities without subjecting their entire enterprise to Federal Reserve supervision.

Insider Threat — A cybersecurity risk originating from individuals within an organization — employees, contractors, or business partners — who misuse their authorized access to systems or data, whether intentionally or through negligence.

Investment Adviser — Under the Investment Advisers Act of 1940, any person or firm that, for compensation, is in the business of advising others about securities. Investment advisers with more than $110 million in assets under management must register with the SEC and are subject to a fiduciary duty to their clients.

K

KYC (Know Your Customer) — The collective term for the obligations financial institutions have to verify the identity of their customers and understand the nature of their activities. KYC requirements exist to help governments track illegal money and ensure institutions can identify and report suspicious activity. KYC encompasses customer identification, customer due diligence, and ongoing monitoring.

L

Liveness Detection — A biometric verification technique used to confirm that the person submitting an identity verification is physically present rather than submitting a photograph or recording. Used in digital KYC onboarding to reduce identity fraud.

M

Malware — Software designed to disrupt, damage, or gain unauthorized access to a computer system. Malware includes viruses, trojans, ransomware, keyloggers, and rootkits. Financial institutions are frequent targets of malware attacks designed to steal credentials or funds.

MFA (Multi-Factor Authentication) — A security mechanism that requires users to provide two or more verification factors to gain access to a system. Factors are typically something you know (password), something you have (a device), and something you are (biometric). MFA significantly reduces the risk of unauthorized account access.

Model Drift — The degradation of an AI model's accuracy or predictive power over time as the real-world conditions on which the model was trained change. A fraud detection model trained on pre-pandemic transaction patterns may underperform when consumer behavior shifts significantly.

Model Governance — The set of policies, processes, and controls a firm uses to develop, validate, deploy, monitor, and retire AI models. Effective model governance includes independent validation before deployment and ongoing monitoring for drift and bias.

Money Services Business (MSB) — A category under the Bank Secrecy Act that includes currency exchangers, check cashers, money transmitters, and virtual currency exchangers and administrators. MSBs must register with FinCEN and comply with AML program, customer identification, and suspicious activity reporting requirements.

Money Transmitter — A business that transmits money on behalf of customers. Money transmitters are classified as money services businesses and must obtain state money transmitter licenses in each state where they operate, in addition to complying with federal BSA requirements.

N

Nation-State Threat — Cyberattacks conducted by or on behalf of a foreign government, typically targeting critical infrastructure, financial systems, or sensitive data. Nation-state actors are among the most sophisticated and well-resourced threat actors in cybersecurity.

NIST Cybersecurity Framework — A voluntary framework developed by the National Institute of Standards and Technology that provides a structured approach to managing cybersecurity risk. The framework is organized around five functions: Identify, Protect, Detect, Respond, and Recover.

No-Action Letter — A formal statement from a regulatory agency, most commonly the SEC or CFPB, indicating that the agency does not intend to recommend enforcement action against a company for a specific product or activity described in the company's request. No-Action Letters provide regulatory certainty for firms operating in gray areas.

O

OCC (Office of the Comptroller of the Currency) — The federal agency that charters, regulates, and supervises national banks and federal savings associations. The OCC has proposed a special purpose national bank charter for FinTech companies that would provide a federal regulatory framework without requiring the charter holder to accept deposits.

OFAC (Office of Foreign Assets Control) — The U.S. Treasury agency that administers and enforces economic and trade sanctions. Financial institutions must screen customers against OFAC's list of sanctioned individuals, entities, and countries at onboarding and on an ongoing basis.

Open Banking — A framework in which financial institutions provide third-party providers with access to consumer financial data through APIs, with the consumer's consent. The CFPB's Section 1033 rule, finalized in 2024, establishes open banking data rights in the United States.

P

Penetration Testing — Authorized simulated cyberattacks conducted by security professionals to identify vulnerabilities in a system before malicious actors can exploit them. Also called ethical hacking. Results inform security improvements and are a key component of proactive cybersecurity programs.

PEP (Politically Exposed Person) — An individual who holds or has held a prominent public position — including heads of state, senior government officials, military officers, and senior executives of state-owned enterprises — and their close family members and associates. PEPs are considered higher-risk for money laundering and corruption and require enhanced due diligence.

Phishing — A social engineering attack in which the attacker impersonates a trusted entity — a bank, employer, or government agency — to trick the target into revealing credentials, clicking malicious links, or transferring funds. Spear phishing targets specific individuals with personalized deception.

Preventative Controls — Internal controls designed to stop problems from occurring in the first place, before harm can result. Examples include access controls that limit who can authorize transactions, segregation of duties that requires multiple approvals, and encryption that prevents unauthorized data access.

Project Catalyst — The CFPB's initiative for encouraging consumer-friendly innovation in financial services, through which the bureau conducts research on emerging FinTech products, engages with innovators, and issues guidance to help companies understand how consumer protection requirements apply to novel products.

Proof of Work — A consensus mechanism used by Bitcoin in which participants (miners) compete to solve computationally intensive puzzles to validate transactions and add blocks to the blockchain. Proof of work is energy-intensive and has been criticized for its environmental impact.

Purpose Limitation — The principle that personal data collected for one purpose should not be used for a different, incompatible purpose without the individual's knowledge or consent. A foundational data ethics concept and a core requirement of the GDPR.

R

Ransomware — A type of malware that encrypts a victim's files and demands payment — typically in cryptocurrency — in exchange for the decryption key. Ransomware attacks against financial institutions and healthcare organizations have become increasingly sophisticated and costly.

Reg BI (Regulation Best Interest) — An SEC rule adopted in 2019 that requires broker-dealers to act in the best interest of retail customers when making securities recommendations, while stopping short of imposing the full fiduciary duty that applies to registered investment advisers.

RegTech (Regulatory Technology) — Technology designed to help financial institutions comply with regulatory requirements more efficiently and effectively. RegTech applications include automated compliance monitoring, AML transaction screening, regulatory reporting tools, and AI-powered risk assessment systems.

Regulatory Arbitrage — The practice of structuring a business or product to take advantage of differences in regulatory requirements across jurisdictions or legal categories, to minimize regulatory burden or avoid oversight. A persistent challenge for regulators of global FinTech firms.

Regulatory Sandbox — A controlled environment in which companies can test innovative products and services under relaxed regulatory requirements, supervised by the relevant regulator. Sandboxes allow regulators and innovators to develop appropriate regulatory frameworks collaboratively before full compliance is required.

RIA (Registered Investment Adviser) — A firm or individual registered with the SEC or state securities regulators to provide investment advice for compensation. RIAs are subject to a fiduciary duty to their clients and must file Form ADV disclosing their business practices, fees, and conflicts of interest.

Risk Assessment — The process of identifying potential risks, evaluating their likelihood and potential impact, and prioritizing them for response. A qualitative risk assessment considers both the probability that a risk event will occur and the magnitude of harm if it does.

Robo-Advisor — An automated investment platform that uses algorithms to construct and manage investment portfolios based on information provided by users about their goals, time horizon, and risk tolerance. Robo-advisors are regulated as investment advisers and subject to fiduciary duty.

S

SAR (Suspicious Activity Report) — A report filed with FinCEN when a financial institution identifies a transaction involving $5,000 or more that it knows, suspects, or has reason to suspect involves illegal activity, is designed to evade reporting requirements, or lacks a lawful purpose. SAR filing is protected by a statutory safe harbor.

SEC (Securities and Exchange Commission) — The federal agency responsible for protecting investors, maintaining fair and orderly markets, and facilitating capital formation. The SEC regulates securities offerings, trading platforms, and investment advisers, and has taken the position that many cryptocurrency assets are securities subject to its jurisdiction.

Section 1033 — A provision of the Dodd-Frank Act, implemented through CFPB rulemaking finalized in 2024, that establishes a consumer right to access their financial data in machine-readable form and to authorize third parties to access that data on their behalf. The foundational legal basis for open banking in the United States.

Segregation of Duties — An internal control that divides critical functions among multiple employees so that no single person can both initiate and approve a transaction, reducing the risk of fraud and error. A foundational preventative control in financial services.

Smart Contract — A self-executing program stored on a blockchain that automatically carries out the terms of an agreement when predefined conditions are met, without requiring human intermediation. Smart contracts underlie many DeFi applications and DAOs.

Social Engineering — The manipulation of individuals into divulging confidential information or taking actions that compromise security, by exploiting psychological tendencies such as trust, authority, urgency, or fear rather than technical vulnerabilities.

Stablecoin — A type of cryptocurrency designed to maintain a stable value by pegging it to a reference asset such as the U.S. dollar, a basket of currencies, or a commodity. Stablecoins have attracted significant regulatory attention due to concerns about reserve adequacy and systemic risk.

T

Travel Rule — A Bank Secrecy Act requirement that financial institutions transmitting funds above a certain threshold must pass along identifying information about the originator and beneficiary of the transfer to the receiving institution. Extended to cryptocurrency transfers by FinCEN, creating compliance challenges for transfers involving unhosted wallets.

Two-Factor Authentication (2FA) — A specific application of multi-factor authentication requiring two verification factors. Widely used in consumer financial applications to reduce unauthorized account access. Common implementations include a password combined with a one-time code sent to a registered mobile device.

U

Unhosted Wallet — A cryptocurrency wallet directly controlled by the user on their own device, without custody by any regulated exchange or financial institution. Transactions involving unhosted wallets create travel rule compliance challenges for regulated exchanges because there is no receiving institution to transmit identifying information to.

Utilitarianism — An ethical framework that evaluates actions based on their consequences, holding that the morally correct action is the one that produces the greatest good for the greatest number of people. A consequentialist theory associated with philosophers Jeremy Bentham and John Stuart Mill.

V

Virtue Ethics — An ethical framework that focuses on the character of the moral agent rather than rules or consequences, asking what a person of good character would do. Associated with Aristotle. In professional ethics, virtue ethics emphasizes the cultivation of qualities such as integrity, honesty, and prudence.

W

White Hat Hacker — A cybersecurity professional who uses hacking techniques with authorization to identify vulnerabilities in systems before malicious actors can exploit them. White hat hackers conduct penetration testing and participate in bug bounty programs, operating under defined rules of engagement and with explicit legal authorization.

Wirecard — A German payment processing company that collapsed in 2020 after it was revealed that approximately $2.1 billion reported on its balance sheet did not exist. The Wirecard scandal illustrated that formal control structures — including external auditors — can be present but captured, failing to prevent or detect large-scale fraud.

www.ingramcontent.com/pod-product-compliance
Ingram Content Group UK Ltd.
Pitfield, Milton Keynes, MK11 3LW, UK
UKHW051957041026
15590UKWH00009B/2351

9 798995 10013